DO ORD
B 24 JUN 1996
WI
AF843644
WITHDRAWN
BOURNEMOUTH UNIVERSITY LIBRARY

GEOGRAPHIC INFORMATION FROM SPACE

Processing and Applications of Geocoded Satellite Images

WILEY-PRAXIS SERIES IN REMOTE SENSING

***Series Editor:* DAVID SLOGGETT, M.Sc., Ph.D.**

Editor, *EARSeL Newsletter*

Co-Director, Dundee Centre for Coastal Zone Research, UK

Deputy Chairman, Cray Systems, UK

This series aims to bring together some of the world's leading researchers working in the forefront of the analysis and application of remotely sensed data and the infrastructure required to utilize the data on an operational basis. A key theme of the series is monitoring the environment and the development of sustainable practices for its exploitation.

The series makes an important contribution to existing literature encompassing areas such as: theoretical research; data analysis; the infrastructure required to exploit the data; and the application of data derived from satellites, aircraft and *in situ* observations. The series specifically emphasizes research into the interaction of elements of the global ecosystem publishing high-quality material at the forefront of existing knowledge. It also provides unique insights into examples where remotely sensed data is combined with Geographic Information Systems and high-fidelity models of the physical, chemical and biological processes at the heart of our environment to provide operational applications of the remotely sensed data.

Aimed at a wide readership, the books will appeal to professional researchers working in the field of remote sensing, potential users of the information and data derived from the application of remote sensing techniques, and postgraduate and undergraduate students working in the field.

EARTHWATCH: The Climate from Space

John E. Harries, Associate Director, Science Department, Rutherford Appleton Laboratory, Chilton, Didcot, Oxon, UK

REMOTE SENSING AND GEOGRAPHIC INFORMATION SYSTEMS: Geological Mapping, Mineral Exploration and Mining

Christopher A. Legg, United Kingdom Overseas Development Administration, Forest and Land Use Mapping Project, Forest Department, Colombo, Sri Lanka

SATELLITE OCEANOGRAPHY: An Introduction for Oceanographers and Remote Sensing Scientists

Ian S. Robinson, Department of Oceanography, University of Southampton, UK

GEOGRAPHIC INFORMATION FROM SPACE: Processing and Applications of Geocoded Satellite Images

Jonathan Williams, Consultant, Space Division Logica plc, Leatherhead, UK

Forthcoming titles

GLOBAL CHANGE AND REMOTE SENSING

Kirill Ya. Kondratyev, Scientific Research Centre for Ecological Safety, Russian Academy of Sciences, St Petersburg, Russia; A. A. Buznikov, Electrotechnical University of St Petersburg, Russia; O.M. Pokrovsky, Main Geophysical Observatory, St Petersburg, Russia

HIGH LATITUDE CLIMATE AND REMOTE SENSING

Kirill Ya. Kondratyev, Scientific Research Centre for Ecological Safety, Russian Academy of Sciences, St Petersburg, Russia; O.M. Johannessen, Nansen Environment and Remote Sensing Centre, Bergen, Norway; V.V. Melentyev, Nansen International Environmental Remote Sensing Centre, St Petersburg, Russia

GEOGRAPHIC INFORMATION FROM SPACE

Processing and Applications of Geocoded Satellite Images

Jonathan Williams
Consultant, Space Division of Logica plc

JOHN WILEY & SONS
Chichester • New York • Brisbane • Toronto • Singapore

PRAXIS

Published in association with
PRAXIS PUBLISHING
Chichester

Copyright © 1995 Praxis Publishing Ltd
The White House,
Eastergate, Chichester,
West Sussex, PO20 6UR, England

Published in 1995 by
John Wiley & Sons Ltd
in association with Praxis Publishing Ltd

All rights reserved

No part of this book may be reproduced by any means,
or transmitted, or translated into a machine language
without the written permission of the publisher

Wiley Editorial Offices

John Wiley & Sons Ltd, Baffins Lane,
Chichester, West Sussex PO19 1UD, England

John Wiley & Sons, Inc., 605 Third Avenue,
New York, NY 10158-0012, USA

Jacaranda Wiley Ltd, G.P.O. Box 859, Brisbane
Queensland 4001, Australia

John Wiley & Sons (Canada) Ltd, 22 Worcester Road,
Rexdale, Ontario M9W 1L1, Canada

John Wiley & Sons (SEA) Pte Ltd, 37 Jalan Pemimpin 05-04,
Block B, Union Industrial Building, Singapore 2057

A catalogue record for this book is available from the British Library

ISBN 0-471-95833-6

ARCH
BOURNEMOUTH UNIVERSITY LIBRARY
WITHDRAWN
02143070
910.285 WIL

Printed and bound in Great Britain by Hartnolls Ltd, Bodmin

Table of contents

The colour plate section falls between pages 84 and 85.

This book is dedicated to my parents and to my best friend, Judith.

Foreword

Over the past few decades, the science of remote sensing has evolved dramatically. In its early years, remote sensing was oriented towards military applications and its civil use was largely limited to the production of colourful mosaics; they served to brighten up offices but were of little practical use. As techniques for interpretation of remotely sensed data improved, scientists and planners in a wide range of fields began to use classified images in their work. It is only recently, however, that the wider potential of satellites to produce geographic information has started to be recognised.

As Jonathan Williams highlights, satellites (and other spaceborne craft) are only one source of information about the earth. We can produce a more comprehensive and correct understanding of the earth's features by integrating complementary information from different sensors and from other, ground-based sources, such as printed maps, digital geographic databases, and field studies; the combined result is much more than just the sum of its parts. However, in order to produce reliable information out of data from diverse sources, the problems of accurate geocoding and error modelling need to be addressed. Furthermore, raw data from space contain, in addition to the desired 'signal', interference from a variety of sources such as atmospheric conditions, sun-angle, etc. This book provides a thorough and practical discussion of the tools available to ensure that such raw data is properly converted into useful geographic information.

While classical space-derived information, such as that from passive sensors, continues be important in fields such as vegetation monitoring and land-use planning, Williams also explores some of the new applications of space data. Active sensors, such as SAR, show a great potential in expanding the amount of geographic information which can be collected from space. These sensors enlarge satellite coverage to include cloud-covered areas, which were previously difficult to monitor. In addition, these new instruments are showing particular potential in the production of accurate digital elevation models. Elevation information is essential to a wide range of applications and users, but has heretofore been unavailable in many parts of the world.

This is a timely book which should be of great interest to anyone involved in assessing the features of the earth's surface, including both applications scientists and planners. It will help them to optimise the use of space-derived data by producing accurate, integrated geographic information. Its frank assessment of the state of the art of remote sensing should also serve to challenge the research scientists working towards the further evolution of this field.

JEAN-PIERRE CONTZEN
Director General
Joint Research Centre of the European Commission

Introduction

Images taken from earth orbiting satellites have a wide variety of applications ranging from forestry to urban development and are rapidly becoming a significant source of geographic information. The acquisition of geographic information from space is a revolution comparable with the invention of map making, and this book attempts to describe a new way of looking at our planet which promises to touch all of our lives.

Geographic Information From Space addresses the basic problems of transforming satellite data into a precise geographic reference system (geocoding) and how useful geographic information can be derived from geocoded images. It is divided into six chapters:

(1) basic concepts of geocoding—why geocoding is an essential prerequisite for geographic information extraction and how it is achieved;
(2) synthetic aperture radar (SAR) geocoding—the particular characteristics of SAR images that require a different approach to geocoding to be adopted;
(3) geocoded image products—production of mosaics and other map-like products from geocoded images;
(4) global topographic mapping from space—the use of stereoscopic images and other data sources to derive height information in the form of digital elevation models (DEMs);
(5) thematic information from space—the variety of techniques for deriving useful thematic information about the earth's surface from satellite data;
(6) integration of geocoded satellite images in geographic information systems (GISs)— the importance of analysing satellite data in conjunction with other data.

The book is deliberately wide-ranging in scope and is primarily aimed at the growing number of remote sensing professionals and other professional users of geographic information as well as graduate and post-graduate students in geography, remote sensing and geographic information systems. It is hoped that it will also prove of interest to environmentalists, civil engineers, system designers and application users in agriculture, forestry, geology and other areas.

A number of individuals and organisations have kindly donated information and illustrative material to the book and I would like to take this opportunity to thank all of them;

specifically John W. Allen (ERDAS UK Limited, Cambridge, UK), Eric Augenstein (Earth Resource Mapping, San Diego, CA, USA), Julie Clark (Radarsat International, Richmond, BC, Canada), Alden Colvocoresses (United States Geological Survey, Reston, Virginia), Dave Gugan (Laser-scan Limited, Cambridge, UK), John Hughes (GIS World), Professor R Rao (Goverment of India, Department of Space, Bangalore), SPOT Image (Toulouse, France) and a particular thank you to Wade Stewart of MacDonald Dettwiler and Associates (Richmond, BC, Canada).

Finally, I would like to acknowledge the assistance provided (wittingly or unwittingly) by my colleagues in the Remote Sensing Division (Defence Research Agency, Farnborough), Environmental and Space Systems Group (Data Sciences Limited), the UK National Remote Sensing Centre and Serco Space during the five years that it has taken to produce this book, in particular—Graham Atkin, Keith Bagot, Andy Batts, Gary Brisdion, Lisa Curtis, Graham Davison, Mark Elkington, Beryl Ezzard, Duncan Goodacre, Mike Hammond, Don Hardy, Allan Haskell, Paul Howard, Dale and Freda Lodge, Karen Machin, Barry Marsh, Robin Merson, Tony Miles, Keith Muirhead, Alfred Odell, Mike Palmer, Mark Poulter, Lois Proud, Maurice Readhead, Dave Sloggett, Nic Stewart, Ian Thomas and Ruth Wilson.

1

Basic concepts of geocoding

As a source of geographic information digital satellite imagery has certain advantages over traditional sources, such as maps, including:

(1) frequent near-global coverage and availability of large archives of data;
(2) reduced data costs per unit area;
(3) reduced need for expensive surveys and terrestrial ground control;
(4) ability to input directly into digital geoprocessing systems and ready incorporation into Geographic Information Systems (GISs).

It is this last point that provides the true revolutionary nature of geographic information from space, as it becomes clear that the real benefits from satellite remote sensing occur when it can be used in conjunction with other sources of geographic information.

Chapter 1 introduces the basic concepts behind geocoding, that is the transformation of satellite images into a standard map projection so that image features can be accurately located on the earth's surface and so that the image can be compared directly with other sources of geographic information (such as maps). It contains the following sections:

— image geocoding methods: why image geocoding is necessary, the simple techniques used for bulk geocoding and the use of Ground Control Points (GCPs). This is followed by how to calculate the transformation from image to map coordinates, the difference between georeferencing and geocoding and finally how two images can be registered together
— advanced GCP techniques: including how a GCP database is set up, the automatic location of GCPs and how the Global Positioning System (GPS) is increasingly being used to provide accurate ground control
— a model-based approach to geocoding: this section describes the three components (orbit, sensor and ground) of a typical mathematical model for geocoding, followed by the resampling techniques used to assign pixel values to the geocoded image, concluding with methods for validating the accuracy of the geocoded image product
— a SPOT geocoding model: this provides a description of a specific geocoding model used for high-resolution images from the French SPOT satellite and how it is put into practice

— AVHRR geocoding models: this section shows how the geocoding model can be applied to a different type of satellite image, in this case from a sensor flown on a US meteorological satellite.

1.1 IMAGE GEOCODING METHODS

1.1.1 Why geocoding is necessary

In an ideal world images taken from an orbiting satellite would be acquired directly in a usable map projection (UTM for instance) ready for immediate integration with other spatially referenced data. In reality satellite images are subject to a number of effects which introduce distortion into the image. For example:

— the satellite platform experiences changes in orbit and attitude;
— the mechanics of the sensor do not perform to a linear specification;
— the earth is curved and rotates beneath the satellite whilst an image is formed;
— variations in terrain relief cause perspective shifts.

Consequently the image will not be acquired in a geometry which corresponds to a standard map projection. This means that the terrestrial location of any point or feature on an image is not known accurately. Additional problems include the inability to directly relate the image to other spatially referenced data (for example printed map sheets) and the fact that images taken of the same area at different times will not be in register.

Sources of geometric error in satellite images can be divided into three main categories—orbit related, sensor related and ground related; examples are given in table 1.1.

Table 1.1 - Categorisation of geometric errors

Category	Description
Orbit	changes to satellite orbit changes to satellite attitude high-frequency attitude variation (jitter)
Sensor	sensor mechanics non-linearities detector misalignment band-to-band misregistrations variations in swath width and timing
Ground	earth curvature earth rotation local topographic effects

Another possible approach to categorisation of errors (Friedmann *et al.* 1983) is to divide them into high, low and medium frequency errors (where frequency is defined relative to time taken to image a scene) as follows:

low	where effects are much less rapid than scene imaging time, such as the orbit model and the earth geoid;
medium	where effects are of the same order of magnitude as the scene imaging time, for example attitude variations;
high	where effects are much more rapid than scene imaging time, for example sensor scan rate or local earth terrain variations.

To account accurately for all the possible geometric effects requires a precise mathematical model for each component contributing to the distortion—in practice however it may not be possible to formulate an exact model or obtain all the parameters needed (for example detailed satellite ephemeris or sensor engineering data). The use of such models is described in sections 1.3, 1.4 and 1.5.

1.1.2 Bulk geocoding

For many applications the approximate transformation afforded by bulk geocoding may prove sufficient. Initially, the Landsat Thematic Mapper (TM) will be used to illustrate the basic concepts of image geometry, as it is widely used throughout the earth observation community. A summary of the TM technical specification can be seen in table 1.2.

Later in the chapter other sources of satellite images such as SPOT and NOAA-AVHRR are also discussed. Bulk geocoding uses relatively simple models of the satellite orbit and sensor geometry in order to determine the terrestrial location of each pixel. The

Table 1.2. Landsat-TM technical specification

Country of origin	United States
Launch	Landsat 4–16 July 1982 Landsat 5–1 March 1984
Orbit	sun-synchronous (near polar)
Orbit period	99 minutes
Equator crossing	09:45 (local time)
Inclination	98.2°
Altitude	705 km
Coverage	81°N to 81°S
Repeat cycle	16 days
Swath width	185 ×185 km
Pixel size	30 metres (band 6 is 120 metres)
Band 1	0.45–0.52 μm
Band 2	0.52–0.60 μm
Band 3	0.63–0.69 μm
Band 4	0.76–0.90 μm
Band 5	1.55–1.75 μm
Band 6	10.40–12.5 μm
Band 7	2.08–2.35 μm

key components of the orbit model are the satellite position at a given moment in time and its attitude (the direction that the platform is pointing in). The position of the sensor relative to the platform will be known, together with its orientation. These, along with a simple model of the earth's surface (such as a sphere or an ellipsoid) provide the basis of the mathematical model.

One fundamental category of distortion that all images are subject to is that caused by the shape of the earth. For bulk geocoding purposes the earth can be regarded as a sphere or an ellipsoid (more accurate models for the earth's geometry are described in section 1.3). Another terrestrial effect that must be accounted for is the earth rotation effect. This effect is caused by the fact that during the time taken to acquire an image the earth will have rotated relative to the satellite ground track. By considering the relative velocity of the satellite vector to the earth's rotation vector the magnitude of the effect can be calculated and a simple correction can be derived from this to slip the image every few lines; this leads to the familiar parallelogram appearance of Landsat images.

The parameters used as input to the bulk geocoding model are the satellite ephemeris data, and the satellite/sensor engineering data, such as angle of intermediate field of view (IFOV) and sensor geometry. Applying the transform will result in an image in a nominal map projection (although inaccuracies, for example, in-orbit timings will persist through the model into the map projection).

An example of the bulk image processing for Landsat-TM (Clark 1990), originally by NOAA and subsequently by EOSAT, consists of three steps:

— raw data ingested;
— geometric correction;
— user products generated.

The default map projection used for geocoded products is space oblique mercator (SOM) with UTM or polar stereographic on request. SOM was invented by Colvocoresses of the United States Geological Survey in 1974 and is particularly suited for the geometric processing of satellite images (Brooks *et al.* 1981). The SOM is a conformal mapping along the ground track of the satellite and uses as axes the along-track and across-track directions. The geocoded image is resampled from the original image values using a cubic convolution method (see section 1.3.4 for more details).

Bulk geocoding will usually result in typical errors of the order of 1 km for Landsat. It should be noted that most bulk geocoded images are intrinsically very geometrically accurate, a typical length distortion being 0.15% (Westin 1990), and if the relative error offset (due to miscalculation of the origin) is subtracted much smaller absolute errors may be obtained. This can, for example, be performed by using a single ground control point (see section 1.1.3) or overlaying coastline data (see section 1.5).

1.1.3 Ground control point methods

A more accurate geocoded product can be obtained by identifying a network of ground control points (GCPs). A GCP (Davison 1986) is a feature which can be readily identified on both the image to be transformed and the map being used to establish its geographic location. Suitable GCPs will contrast strongly with their background and should not be too large in extent so as to preclude accurate determination of their location. Ideal GCPs include such linear features as motorway junctions, railway crossings and runways. Coastal features such as islands and headlands are suitable as long as there is no noticeable effect due to tidal variation. Similarly, lakes and reservoirs can often be used as long as changes in water level do not effect the apparent location of the GCP. Features subject to seasonal or annual variation should be avoided—for example woodland boundaries are often clearly identifiable on an image but may well not still be in the same location as when the reference map was last updated.

The geographic location of GCPs can be established in various ways depending on the accuracy required and the equipment available. The position in image coordinates is usually determined using an interactive image processing facility (IIPF). The capabilities of an IIPF are very much system dependent but will usually include the abilities to pan and zoom images and perform various image enhancement operations (see chapter 3 for examples). After the operator has read in the image and displayed it on the IIPF the cursor is used to select a sub-image containing a GCP. The sub-image is then displayed on the IIPF and using the cursor a coarse location for the GCP is selected. At this stage it is convenient to magnify the area around the coarse GCP location and use the cursor to select a precise location for GCP. This is repeated until a satisfactory distribution of GCPs has been achieved. The simplest method of determining the map location is to measure by eye (for example using a roamer marked with a map grid) from a standard map. More accurate results may be obtained by using a digitising tablet. This is an electronic device which uses a mouse or puck equipped with crosswires which is moved across the map attached to the surface of the tablet. The puck is placed above the GCP to be measured and the exact location determined, typically by using a fine wire grid embedded in the tablet, or by means of electrical wave phase.

1.1.4 Calculating the transformation matrix

The simplest geometric model used is that the GCPs form a representative sample of the entire image, therefore a transformation matrix determined from them is applicable to the entire image. The usual model assumed is a two-dimensional polynomial. After the set of GCPs has been determined for an image two lists of control point locations will have been determined—the first describes their locations in map coordinates, the second in image coordinates. The transformation matrix describes the equation that relates one set to the other, for example, using a linear polynomial:

$$\boldsymbol{X} = \boldsymbol{M}\boldsymbol{x} + \boldsymbol{c} \tag{1.1}$$

where

$\boldsymbol{X}$ = vector location in image coordinates,

$\boldsymbol{x}$ = vector location in map coordinates,

$\boldsymbol{M}$ = 2×2 transformation matrix to be determined,

$\boldsymbol{c}$ = constant offset vector,

Since there may be typically 50 pairs of GCPs the equation for $\boldsymbol{M}$ is over-determined, that is there is more information provided than is needed to solve the equation. As it is not therefore possible to solve the equation exactly a least-squares technique (Ford and

Zanelli 1985) is used to determine the best-fit transformation. The best-fit transformation is defined to be that which minimises the sum of squares of the residual GCP vectors. The residual vector is defined to be the difference between a GCP's true image position and that given by the transformation, that is:

$$\boldsymbol{R} = \boldsymbol{X} - (\boldsymbol{M}\boldsymbol{x} + \boldsymbol{c}). \tag{1.2}$$

The resultant matrix, $\boldsymbol{M}$, is then stored for use in the actual transformation. It is possible to use quadratic, cubic or even higher-order polynomials as the basis for the model but care must be taken—although such polynomials may seem better because they fit the control point set, increased instability can lead to the introduction of artificial errors at intermediate points and thus lead to unwanted image distortions. This effect can be seen in one dimension in figure 1.1.

Another approach to calculating the transformation matrix (Derenyi and Pollock 1990) is to display a roughly transformed image on an IIPF together with a digital map in the desired projection. This can then be used to incrementally refine the transformation matrix in areas where the fit appears to be poor (for example where distribution of GCPs is sparse or local terrain-height variations have induced distortion). Although it is usually more convenient to scan a paper map and import it in raster format better results may be obtained if vector map data is used. This is because:

(1) continued iteration of the transformation and resampling process will result in loss of image information;
(2) it is computationally less expensive to transform map data.

In this case the vector data are subjected to a chain of elementary transformations (such as scale, rotate, translate, or skew). Once a satisfactory fit has been obtained the inverse transformation is applied to the image data itself.

1.1.5 Applying the transformation matrix

Image georeferencing

There are two generally used approaches to the transformation of satellite date (Emery *et al.* 1989):

— transform a map, or set of map points, to fit the image;
— transform the image to the map projection.

The first method, known as georeferencing (or direct image referencing), is shown in figure 1.2.

The advantage of this method is that the data to be transformed (such as digitised maps or grids) are usually stored in vector format which is much more compact than the raster format used to store image data. This means that the volume of transformed data and the amount of processing required is reduced. Another advantage is that there is no loss of information due to resampling (see section 1.3.4). The disadvantage is that the image cannot be directly compared to other sources of spatially referenced data or used for multitemporal work. Thus, this method is generally only useful for single images rather than multi-image sets.

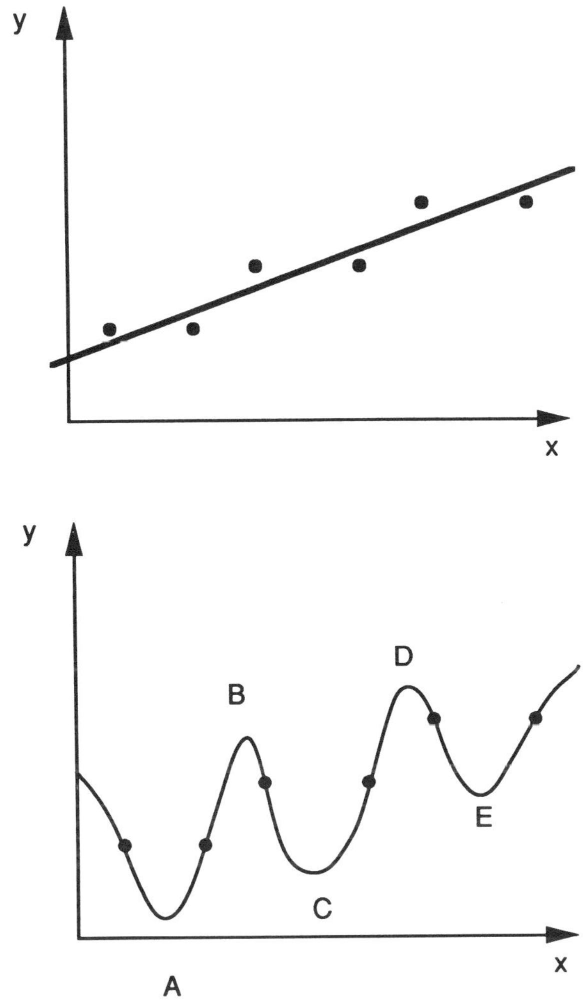

Fig. 1.1 Effect of polynomial distortion between control points a) a linear function provides the best fit straight line between control points but does not pass through any of them b) a higher-order polynomial passes through all the control points exactly but does not provide a good fit at intermediate points (marked A to E).

One example of the applications of georeferencing was a project to locate the position of two Scottish islands (Benny 1985). The islands, Sule Skerry and Stack Skerry, lie 50 km north of the Scottish mainland and to the west of the larger Orkney islands. The islands were surveyed in the nineteenth century by the Admiralty with a positional error of a few hundred metres. The locations were more accurately triangulated by the Ordnance Survey in 1897, but it was not considered worthwhile to incorporate the more accurate positions in the Admiralty charts. The investigation was carried out in 1983 as part of a project which involved the determination of GCPs for the whole of Scotland. Using the above

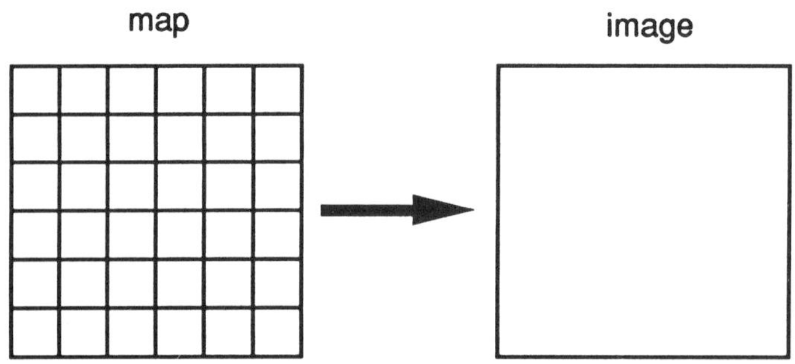

Fig. 1.2. Direct image referencing—the map grid is transformed to fit a particular image.

methods to calculate the transformation on the GCP set enabled the islands' actual position to be georeferenced to Landsat MSS data. This showed substantial errors with their recorded map position and subsequent investigation revealed close agreement with the unincorporated 1897 triangulation data.

Image geocoding

The second and more common method involves transforming the image to the map projection and is known as geocoding. The transformation matrix used with geocoding is the inverse of that used for georeferencing, since we are now transforming the image to fit the map (see figure 1.3).

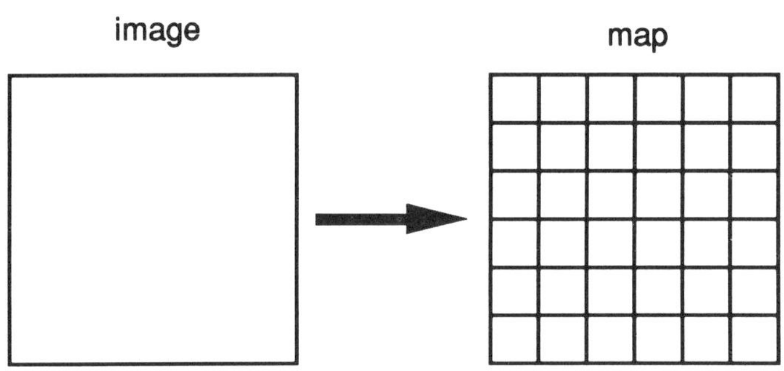

Fig. 1.3. Image geocoding—the image is transformed to fit the map grid.

It is usual to produce a transformed image on a regular grid. This makes it easier to calculate map locations and combine with other spatially referenced datasets. The grid spacing for a particular data source is normally about the same size (or slightly smaller) than the input pixel size. It may also be advantageous to use a standard scale to allow different sources of data to be overlaid. Typical grid spacing for various data sources are shown in table 1.3.

Table 1.3. Typical grid spacings for image geocoding

Satellite	Sensor	Grid spacing (m)
NOAA	AVHRR	1000
Landsat	MSS	100
		50
	TM	30
		25
SPOT	XS	20
	PA	12.5
		10

The extent of the transformed image can be specified in various ways (see figure 1.4), for example

(1) entire input image (in this case the output will usually be padded with blank values to produce a rectangular image);
(2) the largest rectangle on the map that fits inside the input image;
(3) specifying a sub-image in map coordinates.

The actual transformation algorithm proceeds as follows—first a grid in the required map projection is set up at a suitable (regular) spacing. Then for each line in the output grid, and for each point in that line the inverse transformation is applied to that point to calculate its position in the original image (if the transformed point is outside the original image then the output pixel is set to a blank value). If the point lies within the boundary of the original image then the nearest pixel in the original image is calculated and the output pixel set to the value of that pixel. The resampling step of calculating the nearest input pixel to the transformed image point is necessary because the transformation will (in general) not correspond to an exact integer location. This method of resampling is known as 'nearest neighbour'; it is simple to implement but not particularly accurate. Expressed mathematically:

$$O(x, y) = I(X', Y') \qquad (1.3)$$

where

$O =$ value in output (geocoded) image,

$I =$ value in input image,

$(x, y) =$ location in map coordinates,

$(X, Y) = f(x, y)$,

$f =$ transformation function,

$X' =$ int $(X + 0.5)$,

$Y' =$ int $(Y + 0.5)$.

More sophisticated resampling techniques are described in section 1.3.4.

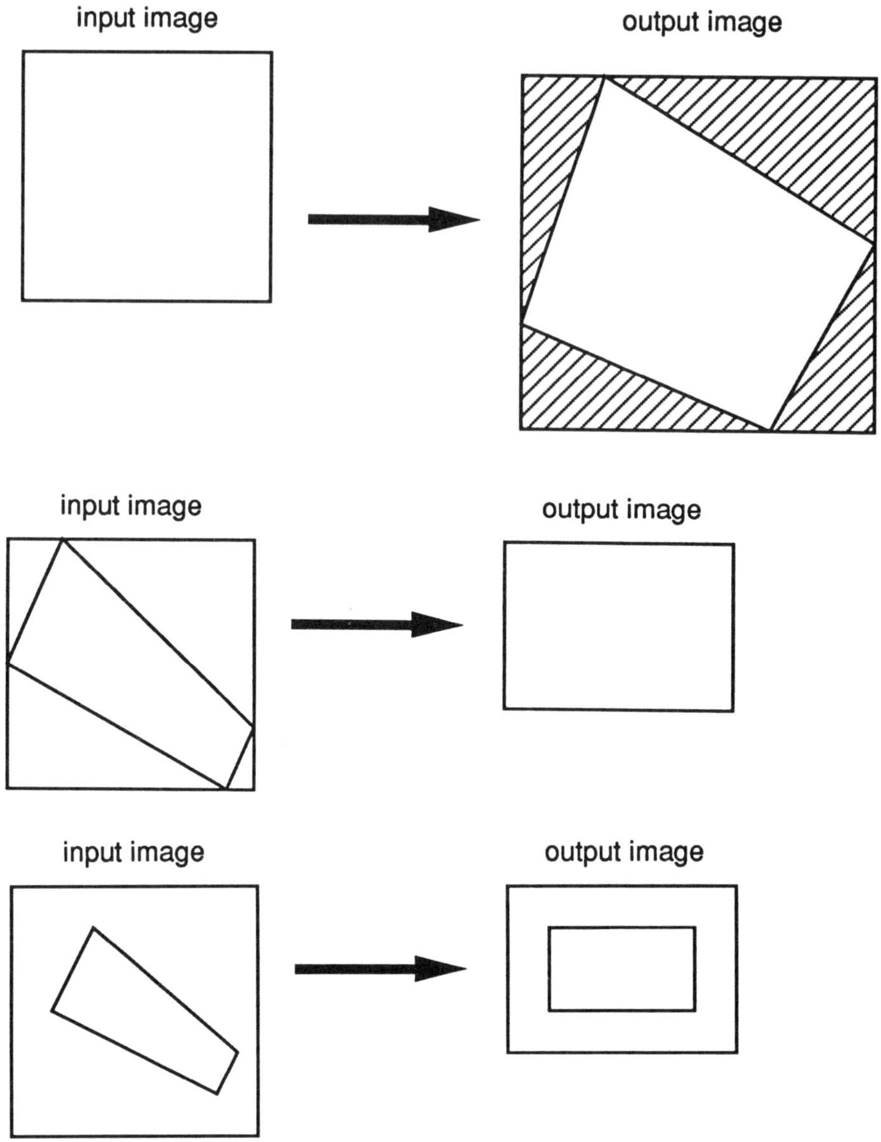

Fig. 1.4. Methods for specifying the extent of the output image: (a) the entire input image is selected and the output is padded with blanks; (b) the largest map rectangle that entirely fits the input image is selected; (c) a rectangular sub-image is specified in map coordinates

1.1.6 Image registration

An alternative transformation method is image registration. This is usually of value when the images are to be compared directly (for example multitemporal analysis of changes), or when a reference map is unavailable. Faced with the difficulty of deriving an acceptable mathematical model the use of unmodelled transformations of Landsat images has become widespread. The advantage of such methods is that they require little image-specific information (such as satellite orbital information). The registration algorithm is based on using tie points which can be identified and located in both images. Their absolute map reference does not have to known, thus this method can be used when local maps are unobtainable, or unreliable. One image is selected as the base image—the second image is registered to this. The actual registration process is carried out using an IIPF. For the purposes of image registration the display is often divided into four windows. Two of the windows are used to store an overview of the images to be registered. Since a typical SPOT image may be 6000×6000 pixels the images are heavily sub-sampled before display.

The image registration process commences by an operator selecting an area that is likely to contain a suitable control point. The area is then copied at full resolution to a third window. A similar operation is performed for the second image. A tie point that can be identified in both image windows is selected. Facilities to assist selection of tie points include contrast stretching and edge enhancement (see section 3.2 for details), as well as the ability to progressively zoom-in to control points to enable the operator to accurately place a cursor on the tie point location. After a few points have been selected in this manner a fairly accurate matrix can be calculated. This allows the program to predict the location in the second image given the cursor position in the base image to a few pixels. This greatly helps the operator select the next control point. When sufficient control points have been selected the transformation matrix can be calculated in the normal way. Transformation and resampling proceed using the method described in Section 1.1.5.

1.2 ADVANCED GCP TECHNIQUES

1.2.1 A GCP database

Since the same GCPs can be utilised for any images covering similar areas it is possible to eliminate a lot of the tedious effort involved by setting up a database of GCP locations in map coordinates. Table 1.4 shows an extract from such a database used by the UK National Remote Sensing Centre, showing a selection of GCPs from Kenya. The database identifies for each GCP the name of the country, the map sheet number, GCP sequence number on that map, easting, northing and elevation in metres and finally a brief descriptive text.

Typical database operations include addition, deletion and modification of GCP details (for example entering height value for a previously defined point). Another useful check is to determine points which are very close to a GCP being entered and which may simply be that point subject to a certain measurement error. For example a warning may be issued when a candidate GCP is within 100 metres of a GCP that has been previously entered on the database. The approximate locations stored in an image header, or calculated by bulk transformation, provide an initial estimate of the area covered by a particular image. Allowing a margin of error of a few kilometres, depending on the images used, produces a geographical search area in which the GCPs for the image can be expected to lie. The database query facilities can be used to produce a candidate set of GCPs. This can be used to display the rough locations on an IIPF, which can then be fine-tuned interactively by the operator.

Table 1.4. Extract from a GCP database (based on Davison 1986)

			Grid reference		Height	
Country	Map	GCP	East	North	(m)	Description
EAK	1742	1	383092	9774976	780	Bend in R Nr Yatta Plateau
EAK	1742	2	385101	9760470	750	Bend in R Athi Nr Milembwa
EAK	1742	3	378209	9764706	800	Bend in R Kikku Nr Kamisitu
EAK	1742	4	368383	9765682	850	River junction Nr Kiteet
EAK	1742	6	384292	9767628	700	Bend in river Nr Yatta
EAK	1744	1	367354	9749532	900	Bridge on Kiboko Ranch
EAK	1744	2	479067	9749842	1200	Top point on Kilema Range
EAK	1744	4	378545	9728230	1100	Umani
EAK	1744	5	386676	9733640	890	Bend in road through plantation
EAK	1744	6	387490	9738302	894	Bend in road Kalungu
EAK	1751	2	407824	9771752	650	Bend in river Nr Ikutha
EAK	1751	4	396627	9755238	725	Bend in river Nr Athi new bridge

1.2.2 Automatic selection of GCPs

If GCPs have to be located for each new image, even using a database approach to speed the operation can still entail a very labour-intensive task. It can be greatly improved by the use of automatic relocation techniques. To do this a library of GCP 'chips' extracted from actual images is set up. A typical method of using GCPs which has proved successful for use with Landsat-TM imagery (Benny 1983) is as follows. The standard satellite information is used to produce an initial transformation matrix—however, as with most bulk geocoding techniques, the result may well be intrinsically very accurate geometrically but still require location to the correct origin. Typically a single GCP will suffice to complete the initial matrix. Based on this the GCP library is searched for all GCP chips that can reasonably be expected to lie on the image. The set retrieved is sorted in order of their distance from the first point. The initial matrix is used to provide the best guess for the location for the second GCP and a spiral search commenced at this point until the real best-fit location is found. This process is continued for all subsequent GCPs with at each stage the true locations fed back into fine-tuning the transformation matrix. To safeguard against spurious matches two constraints may be placed on the search criteria for each GCP:

(1) the minimum correlation value that is regarded as a successful match;
(2) the maximum distance from the expected location that a successful match can be accepted (this value can be adaptive–decreasing in value as the matrix iteratively becomes more accurate).

The algorithm for GCP chip location consists of the following steps:

(1) using the IIPF cursor locate a central GCP on the image;
(2) retrieve the chip corresponding to this GCP;
(3) for each point (x, y) in a search area around this initial location calculate the correlation between chip and image at (x, y);
(4) set the GCP location to that of the correlation maximum in the search area;
(5) use the location of the initial GCP to define the origin of the transformation;
(6) calculate the approximate image transformation using known parameters (such as spacecraft height and heading);
(7) retrieve a candidate set of GCPs that should appear on the image and sort into ascending distance from the origin.

Repeat steps 8 to 11 for each GCP in the set:

(8) using the current approximate image transformation estimate the location of the GCP in the image;
(9) for each point (x, y) in the search area calculate the correlation between chip and image at (x, y);
(10) if the maximum correlation is less than threshold mark the GCP as 'not found';
(11) otherwise set the GCP location to that of the correlation maximum in the search area and recalculate the transformation matrix.

For cloud-free summer images, typically over 95% of the GCPs for a Landsat-TM image can be relocated by this method. For more complex images, such as SPOT, simple correlation techniques such as those used for Landsat can be unsuitable for chip matching. Landsat images with their fixed nadir viewing ensure that chips from different passes have the same geometry. Off-nadir imagery from SPOT has a variable viewing-angle which necessitates the comparison of chips with different geometries, for example using image matching techniques (see section 4.2). It is also possible to synthesise image chips from data of a smaller pixel size, for example making use of Landsat MSS to produce AVHRR chips, or aerial photography to produce Landsat chips. As the need for multisensor registration becomes more widespread greater use of geographic models of image features is likely. These could, for example, consist of digital models of a GCP's location, shape, edges, texture, background, and seasonal variation. The model for a particular GCP can then be used to simulate the appearance of an image chip for a particular sensor geometry.

1.2.3 Use of global positioning system

The United States Global Positioning System (GPS) is a satellite-based high-accuracy positioning system which can be accessed from anywhere on the earth's surface (Mack 1992). The complete system consists of 21 satellites (plus three back-up platforms) which can be used to locate the terrestrial position of a GPS receiver. GPS is increasingly being used in conjunction with satellite images to provide a rapid and accurate source of mapping information. Examples of projects that have been conducted using a combination of GPS and remote sensing (Hough 1992) include:

(1) vegetation mapping of the Ituri rain forest in Zaire by the University of Utah;
(2) production of image maps from Landsat-TM for use by allied forces during the Iraqi occupation of Kuwait;
(3) detection of gold deposits by the Geological Survey of Japan (jointly with the Research and Development Centre for Geotechnology of Indonesia) on the Indonesian island of Lomblen.

Originally designed as a US Department of Defense system the accuracy has been deliberately degraded in selective availability (SA) mode for civilian use. In this mode the RMS accuracy attainable with a single receiver is in the order of 50 m. To obtain higher accuracy GPS can be used in differential mode by comparing the position of a GPS receiver in the field with the known position of a GPS base station. This can result in an accuracy more of the order of 5 metres, which enables features suitable for use as GCPs to be located to sub-pixel accuracy for most commonly used types of satellite images. To obtain a positional fix (Sharun 1993) signals from at least three different satellites must be obtainable; in practice this means that they must be at least 15 degrees above the horizon. Signals from three satellites enable the horizontal position of the GPS receiver to be determined; if four satellites are available then a full three-dimensional fix can be obtained.

1.3 A MODEL-BASED APPROACH TO GEOCODING

To produce a highly accurate geocoded image it is essential to adopt a model-based approach to geocoding. This allows the setting up and manipulation of exact mathematical equations which describe every aspect of the image geometry. This allows not only a controlled and methodical means of producing a geometrically accurate product but allows errors to be tracked to particular parts of the model which can subsequently be refined. The sequence of steps preformed in a typical geocoding operation are as follows:

(1) set up the components of the mathematical model and calculate the transformation between image (pixel) coordinates and the desired map projection by introducing parametric data to the mathematical model;
(2) resample the original image data onto the output grid (using the calculated transformation) to produce a geocoded image;
(3) validate the model, that is determine that the transformation that has been calculated is physically reasonable.

As the sources of image distortion can be divided into orbit related, sensor related and ground related errors; so the components of a typical geocoding model can be divided into three sub-models:

— orbit model;
— sensor model;
— ground model.

1.3.1 Orbit model

The purpose of the orbit model is to be able to locate the instantaneous position of the satellite sensor (at a time that a particular pixel was imaged) with respect to a geocentric coordinate system—it is also important to be able to determine which way the satellite is pointing (its attitude). Thus the orbit model can be conveniently broken down into two component models:

(1) orbit position model—the (x, y, z) position of a reference point on the satellite structure as a function of time;
(2) orbit attitude model—the location of the three local orbital reference axes (pitch, roll and yaw), also as a function of time, coupled with the calculation of an attitude reference vector of a sensor with respect to these axes.

The conventional definition of the three orbital axes (derived originally from aircraft terminology) is as follows:

- yaw —a vector from the earth's centre to the instantaneous position of the satellite's centre;
- roll —a vector in the orbit plane (perpendicular to the yaw vector) in the direction of the satellite's motion;
- pitch —a vector (also in the orbit plane) perpendicular to the yaw and roll vectors to complete a right-handed system.

The attitude reference system is another right-handed orthogonal set, this time fixed in the satellite (with origin the satellite centre), used to describe the three components of the satellite's rotation about its instantaneous centre. A specific example of derivation of an orbit model can be found in section 1.4.

1.3.2 Sensor model

The sensor model can be one of the most complex sub-models to produce involving the geometry, mechanics and optics of the sensor itself. It is necessary to consider each satellite sensor individually and to model it accordingly, for example (O'Brien and Prata 1990) the ERS-1 Along-track Scanning Radiometer (ATSR) has a novel conical scanning system that views a point on the earth's surface once on a forward scan and once on a backward scan. Thus the ATSR sensor model must describe the rotation scan mirror which traces a circle of 450 km radius on the earth's surface in 0.15 seconds. With a scanning system such as Landsat (MSS or TM) the model will concentrate on the mechanics of the sensor—the rate of scan for example. The aim is to be able to calculate the instantaneous look direction of each detector within the attitude reference. With push-broom sensors there is no corresponding mechanics as the motion of the sensor is provided by the motion of the satellite itself. When modelling a sensor it is generally not possible to completely separate the radiometric properties of an image from its geometric ones, thus there will always be some 'cross-talk' from the geometric aspects of a sensor model to its radiometric ones. One of the most important aspects of any sensor model is its 'resolution'—this term is often used interchangeably with 'pixel size' or 'intermediate field of view' (IFOV)—but it is important to distinguish between them.

The intermediate field of view is determined by the sensor geometry and mechanics and describes the portion of the earth's surface that can be seen at any one instant. The IFOV may not be a constant value, for example due to changes in platform height and orientation. With scanning sensors the IFOV is commonly determined by a fixed angle of view; this can also lead to variable sizes of IFOV, for example with AVHRR images the very large swath width, coupled with the effect of the earth's curvature means that extreme off-nadir IFOV is significantly greater than central IFOV.

The pixel size is the terrestrial area represented by a single pixel in the satellite image; this may not necessarily be the same as the IFOV, as many images will be deliberately resampled to have a constant, uniform pixel size. Finally, the definition of resolution is a more qualitative concept, the basic definition being based on the smallest size target that a sensor is capable of detecting. For terrestrial targets, however, their detectability is very much dependent on their geometry, texture and contrast with their surroundings (for example bright objects that are smaller than pixel size may have sufficient contrast to significantly alter the pixel value above the surrounding background, and hence to become detectable).

It is also important to consider the true spatial extent of the information 'encoded' into a pixel—in reality sensors will not provide a simple block response exactly covering a square pixel area on the ground. Sensor response is actually characterised by a two-dimensional point spread function (PSF) centred on a pixel but extending beyond it. For Landsat (Wilson 1988) the PSF can be represented as the product of two one-dimensional line spread functions (LSFs):

$$\mathrm{PSF}(x, y) = \mathrm{LSF}(x) \times \mathrm{LSF}(y) \tag{1.4}$$

where

$\mathrm{PSF}(x, y) =$ two-dimensional point spread function,

$\mathrm{LSF}(x) =$ one-dimensional across-track line spread function,

$\mathrm{LSF}(y) =$ one-dimensional along-track line spread function.

A diagram of the across-track LSF (x) can be seen in figure 1.5; the function is largely determined by the combined effect of the sensor optics, electronics and detectors.

The along-track function $\mathrm{LSF}(y)$ is similar but the effect of electronics is negligible. The PSF will vary from sensor to sensor and a knowledge of this function will contribute greatly to accurate modelling of aspects such as radiometric calibration and design of resampling kernels.

1.3.3 Ground model

The ground model consists of two components:

earth model – a description of the overall shape of the earth;

terrain model – a more detailed model of local scene topography.

The equation for a spherical earth model is given by:

$$x^2 + y^2 + z^2 = R_e^2 \tag{1.5}$$

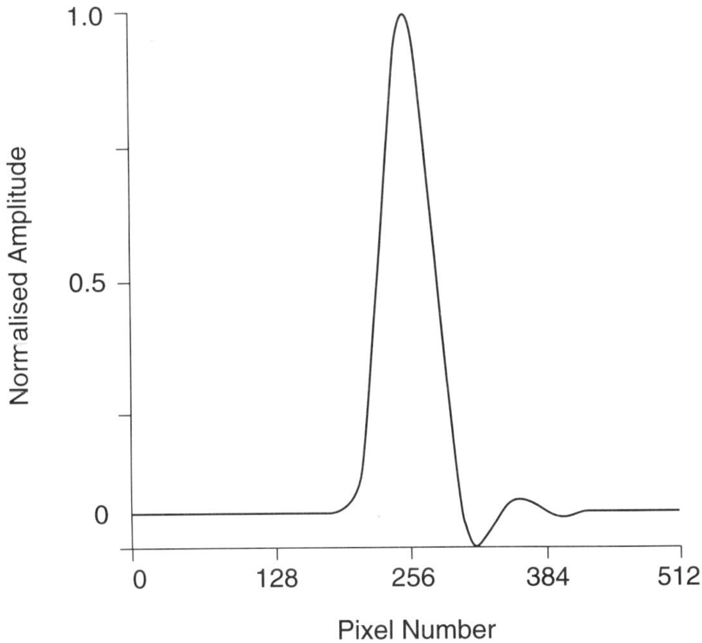

Fig. 1.5. Across-track line spread function for the Landsat-4 primary focal plane detectors (from Wilson 1988).

where

$(x, y, z) =$ location of a point to the earth's surface,

$R_e =$ mean equatorial radius (6378.138 km).

A more accurate model (Curlander *et al.* 1987) of the earth as an oblate ellipsoid is given by the following equation:

$$\left(\frac{x}{R_e}\right)^2 + \left(\frac{y}{R_e}\right)^2 + \left(\frac{z}{R_p}\right)^2 = 1 \tag{1.6}$$

where

$(x, y, z) =$ location of a point on the earth's surface,

$R_e =$ mean equatorial radius (6378.138 km) + average scene elevation,

$R_p = (1 - 1/f)R_e$,

$f =$ flattening factor ($f = 298.255$).

To produce a geocoded image fully corrected for terrain effects (an orthoimage) needs a detailed model of the image geometry and the local terrain. This can only really be provided by a detailed digital elevation model (DEM) of the earth's surface. Ideally this should be of a similar grid-spacing to the image pixel size (for example 30 metres for Landsat-TM). Provision of DEMs is not very good for most of the earth's surface. Thus the most suitable approach may well be to derive the DEM required from the satellite image itself as part of a stereo-pair (this technique is described in more detail in chapter 4). This is one example of the synergy between satellite images and 'world models'—a topic returned to under the heading of geographic information systems (GISs) in Chapter 6.

Although not strictly part of a ground model the use of other terrestrial objects such as GCPs and coastlines is an important part of the geocoding process, having two main applications:

(1) to validate the model, for example by measuring the accuracy of the transformation at a set of independent check points (see section 1.3.5);
(2) to provide a low-frequency offset to a model which typically will account for all the high-frequency geometric aspects but may need 'nudging' into position.

1.3.4 Resampling

Once the geocoding model has been set up, and the transformation calculated, the final step is to resample the values from the original image to points on a regular output grid. The basic requirement is to be able to transform a particular point in (x, y) map coordinates back into the original image coordinates and to estimate what the corresponding image value should be at that point. The simplest method and the one generally used for most bulk geocoded products is to use the value of the nearest pixel in the original image—otherwise known as nearest neighbour interpolation. A more accurate method is bilinear interpolation, whereby a weighted average of the four surrounding pixels is used (see figure 1.6). The algorithm used is defined as follows:

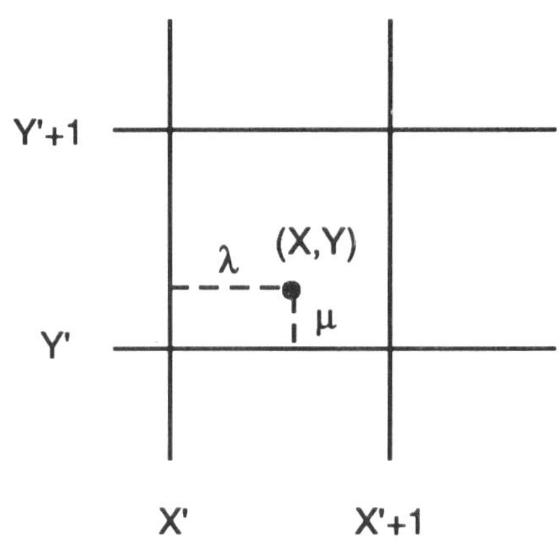

Fig. 1.6. The geometrical basis for bilinear interpolation.

$$O(x, y) = (1-\lambda)(1-\mu)\times I(X', Y') + (1-\lambda)\mu \times I(X', Y'+1) + \lambda(1-\mu)\times I(X'+1, Y') + \lambda\mu \times I(X'+1, Y'+1) \qquad (1.7)$$

where

$$O = \text{value in output (geocoded) image,}$$
$$I = \text{value in input image,}$$
$$(x, y) = \text{location in map coordinates,}$$
$$(X, Y) = f(x, y),$$
$$f = \text{transformation function,}$$
$$X' = \text{int}\,(X),$$
$$Y' = \text{int}\,(Y),$$
$$\lambda = X - X',$$
$$\mu = Y - Y'.$$

Another commonly used algorithm is cubic convolution, which uses a cubic spline to approximate the theoretically ideal interpolator known as the 'sinc' function:

$$\text{sinc}\,(x) = \frac{\sin(\pi x)}{\pi x}. \tag{1.8}$$

As this function has an infinite domain it cannot be computed exactly. Instead (Shlien 1979) it is represented by a piecewise cubic function. One such function (Moik 1980) which is continuous in value and slope is given by:

$$f(x) = \begin{cases} 1 - 2X^2 + X^3 & 0 \le X < 1 \\ 4 - 8X + 5X^2 + X^3 & 1 \le X < 2 \\ 0 & X \ge 2 \end{cases} \tag{1.9}$$

where $X = \text{abs}(x)$.

1.3.5 Validation

The validation of geocoded images is an essential part of the geocoding process—it is vital to be able to have confidence in the accuracy and repeatability of any particular geocoded product. This topic is returned to several times throughout this book. There are many possible definitions and approaches to validation but three complementary approaches to validation can be identified:

blunder detection — trapping of gross errors or physically impossible results, thus ensuring that all products will always be correct to a 'first order';

process validation using the model to produce quality control parameters, for example, generation of quality products and reports as an intrinsic part of the geocoding process. This is often used, for example, as a preliminary check on the accuracy of a transformation matrix before resampling is commenced;

product validation — quality control of the final output product against expected accuracy standards (for example remeasuring selected GCPs against their known map location).

One example of blunder detection described previously is GCP chip location where the search is abandoned if the search area extends too far from the estimated GCP location or if the correlation maximum is below a certain threshold. Process validation is often derived from statistical analysis of the set of GCP locations and the transformation derived from them, however single metrics (such as RMS error) may not be particularly useful. A better approach (Harrison *et al.* 1989) is to produce an entire 'error band' showing the accuracy of the transformation at all locations in the image. Because the transformation equation is over-determined the resultant matrix will not fit the GCPs exactly. One measure of the accuracy of a transformation is to examine the residual error at each GCP. This is determined by applying the transformation to each GCP and comparing it with its known position, using equation (1.2) to determine the residual error. A simple way to examine the residuals is to display them as vectors overlaid on the original image. More precise mathematical analysis is also possible. To provide an unbiased result certain of the original set of GCPs may be assigned as check points—these are not used to calculate the matrix but are used as part of the residual plot, to provide a better idea as to how good the geocoding process is away from the control point network. Other approaches to process validation include analysis of the GCP distribution, transformation function analysis and production of an error budget which ascribes errors to each part of the model used (for example error due to DEM, error due to digitising GCPs from maps). An interesting automatic approach to validation of GCP selection is being developed at Nottingham University (Heard *et al.* 1992). The geometric rectification expert system (GERES) is a prototype expert system which can advise on poor spatial distribution of GCPs, advise the operator which areas are particularly lacking in coverage and point out new candidate GCPs in these areas.

Finally, product validation consists in assessing the overall geometric accuracy of the final product. This can be performed by measuring the location of points on the image against their true location on the map (essentially this is repeating the GCP selection exercise). Other methods include comparison of the image with a geocoded product of known accuracy, either by superimposition on an IIPF or by identification of tie points (the transformation between two images geocoded into the same map space should be identical to the unit matrix). Another example of product validation is a method used for quality assessment of geocoded Meteosat images (Adamson et al. 1988) before dissemination to the meteorological community. Meteosat is a geosynchronous satellite, located over the equator at a longitude of 0 degrees, that provides 2.5 km pixel size visible images and 5 km pixel size infra-red images. In theory a geostationary satellite should produce images that need no rectification (apart from transforming to a convenient map projection) but in practice variations in orbit and attitude, particularly after a satellite manoeuvre, introduce distortions that require removal by geocoding. The validation process automatically calculates the displacement from image features to a library of landmarks (typically 10 x 10 pixels in size) which are mainly water boundary features such as peninsulas or inland lakes. The best landmarks are isolated, symmetrical islands such as

Mallorca. A correlation technique is used to find the best-fit location of landmarks in the actual image and, hence, to validate the geocoding accuracy.

1.4 A SPOT GEOCODING MODEL

1.4.1 SPOT sensor model

One of the most important missions for cartographic applications is the French SPOT mission which commenced with the launch of SPOT 1 from Kourou in French Guiana in February 1986. During its first four years of operation the SPOT satellite acquired 1.5 million images of the planet, approximately 80% of which are usable, and a continuity of service beyond the year 2000 is currently planned. A summary of the SPOT technical specification can be seen in table 1.5.

Table 1.5. SPOT technical specification

Country of orgin	France
Launch	SPOT 1—22 February 1986
	SPOT 2—22 January 1990
Orbit	sun-synchronous (near polar)
Orbit period	101 minutes
Equator crossing	10:30 (local time)
Inclination	98.7°
Altitude	830 km
Coverage	81°N to 81°S
Repeat cycle	26 days
Swath width	60 km (nadir)
Pixel size	10 metres (panchromatic)
	20 metres (multispectral)
Panchromatic	0.51–0.73 μm
Band 1	0.50–0.59 μm
Band 2	0.61–0.68 μm
Band 3	0.79–0.89 μm

The SPOT sensor consists of two high-resolution visible (HRV) instruments capable of resolution to 10 m. SPOT replaces the mechanical scanning mirror systems of earlier satellites (typified by Landsat) with the newer technology of push-broom sensors. The push-broom sensor comprises linear arrays consisting of 1728 charge coupled device (CCD) detectors. The scanning of each line of a SPOT image is achieved by the measurement of the current generated by each CCD detector—four arrays are used to detect each spectral band. Another difference with SPOT is that the sensors are movable, by means of mirrors, in 0.6° steps to a maximum of 27° off-nadir. This allows the sensor to be pointed at a particular target, increasing the potential revisit frequency and providing the capability of acquiring stereo-pairs (see section 4.1.1). One aspect of this mechanism is that unprocessed SPOT images can vary in size, from 60 km × 60 km at nadir, to 60 km × 80 km when fully tilted. The SPOT camera model can be compared with that of an aerial photograph. The single most important factor in the applicability of an aerial photograph is that it is acquired almost instantaneously, that is the camera platform does not move a significant distance in the time it takes to form a photographic image. This leads to a simple optical geometry known as single point perspective where the light rays from every point on the surface to the corresponding point on the photograph all pass through a single point (the perspective point). A SPOT image on the other hand, does not have this single point perspective as it is effectively made up of a collection of one-dimensional images acquired at 1.5 ms intervals. Each of these images is taken by an array of detectors lying in a plane approximately defined as a vertical plane perpendicular to the image track. Note that the image plane is independent of the viewing angle as the geometry of the sensor mirror is defined as a rotation about an axis perpendicular to this plane, thus rotating the mirror simply causes the imaging geometry to change within the plane—it does not alter the geometry of the plane itself. As described previously, however, the satellite orbit and attitude does change significantly as the entire image is acquired, the envelope of the image geometry being defined by the continuous variation in the visible planes of each image line. Thus the image geometry can be regarded as being defined by a perspective locus given by the instantaneous perspective centre of each line.

For an image with single point perspective six parameters are needed to determine its exterior orientation, namely its three-dimensional position (x, y, z) and the three angles that describe its attitude (ω, ρ, κ). These parameters only need to be determined once per photograph. In order to be able to process 'dynamic imagery' it must be possible to model the continual variation of the instantaneous values of these parameters. For this purpose a good orbit model is essential.

1.4.2 SPOT orbit model

The SPOT image header contains detailed ephemeris data describing the motion of the satellite. The most important in terms of the SPOT geocoding model are:

- — satellite position;
- — three-dimensional velocity vector;
- — attitude changes of sensor;
- — time when the centre of scan-line was imaged.

The attitude control system (Westin 1990) is designed to keep the satellite within 0.15 degrees of a nominal reference. The SPOT orbit model can be derived by standard methods of orbital mechanics using the ephemeris data available for a particular orbit. The position–velocity data supplied with a SPOT image give the instantaneous position and velocity of the spacecraft at one-minute intervals (the attitude drift rates are recorded in the satellite header eight times every second and they describe how the angular velocity of the satellite about its three axes varies with time). Whilst not sufficient to provide instantaneous information for each line in an image (an entire SPOT image is acquired in nine seconds) the orbit and attitude of the spacecraft is generally stable enough for accurate interpolation between the one-minute data. This leads to two possible models to determine the relationship between the local orbital reference and the attitude reference:

(1) assume that the relationship between the two is constant during the acquisition of a particular scene;
(2) estimate the instantaneous relationship using satellite attitude drift rates.

$a =$ half major axis,

$e =$ eccentricity,

$i =$ inclination,

$\Omega =$ right ascension of the ascending node,

$\omega =$ argument of the perigee,

$M =$ mean anomaly.

For satellite images it is not necessary to model all these parameters accurately, for example during the acquisition of a single image it is sufficient to model the orbit as a plane circular one. The primary parameters are derived from the satellite ephemeris data down-linked with the image. It is also necessary to model the satellite attitude which is given approximately by the equations:

$$\begin{aligned} &\text{roll} && \omega(t) = \omega_0 + \Delta\omega(t) \\ &\text{pitch} && \rho(t) = \rho_0 + \Delta\rho(t) \\ &\text{yaw} && \kappa(t) = \kappa_0 + \Delta\kappa(t) \end{aligned} \tag{1.10}$$

where

$(\omega_0, \rho_0, \kappa_0) =$ initial attitude,

$\Delta\omega(t) =$ change in roll angle at time t (similarly for pitch and yaw).

It can be shown that none of the above approximations results in an error of greater than 0.2 metres which is more than acceptable for geocoding purposes.

1.4.3 Ground model

Another effect that must be accounted for in the precision geocoding of satellite images is the effect of the earth's terrain on the image geometry. As can be seen in figure 1.7, because an optical scanner (or a push-broom sensor) essentially measures angles, an object extending above the surface of the earth will cause an erroneous position to be recorded.

Note that this effect depends on the across-track distance from the satellite nadir, thus it will be zero for points imaged directly at nadir and be most pronounced for targets imaged at large off-nadir angles such as produced by the steerable SPOT sensor. This effect is such that a 200 metre high object causes a relief displacement of nearly one pixel at the end of a Landsat-TM scanline (see table 1.6) the effect being even more pronounced for SPOT images with their off-nadir viewing angle.

With high-resolution off-nadir sensors it is possible to obtain images of the same area with a vast range of viewing angles and a corresponding plethora of sun–target–sensor geometries. One immediate consequence is that of terrain relief—given that the satellite

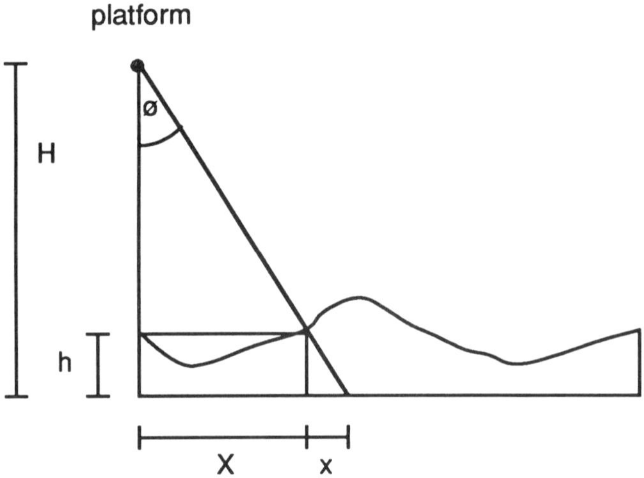

Fig. 1.7. The effect of terrain on image geometry—the true horizontal location of a point at height *h* above the earth's surface (viewed at angle ϕ from a satellite orbiting at height *H*) is *X*, however it actually appears displaced by an additional distance *x*.

Table 1.6. Effect of relief on pixel displacement (adapted from Binnie and Colvocoresses 1987)

	Horizontal distance from nadir (km)		
Elevation (m)	10	50	100
100	1.6	7.9	15.8
200	3.2	15.8	32.0
500	7.9	39.5	79.0
1000	15.8	79.0	158.0

Note: values of displacements are in metres, for example an object 200 metres high, 50 km from nadir causes a displacement of 15.8 metres.

altitude remains constant the apparent displacement is a combined function of the elevation of the point being imaged and its distance from the satellite nadir. This has two effects:

(1) the displacement is independent of pixel size—thus a relief displacement of 30 metres will produce an apparent shift of one pixel on a Landsat-TM image but three on a SPOT-PA image;
(2) off-nadir images are far more effected by relief displacement.

Depending on the accuracy required the relief effect can be modelled by a flat earth model (usually sufficient for narrow swath widths) or a curved earth (normally used for the more off-nadir images). To produce geocoded images fully corrected for the relief effect (orthoimages) the relief correction must be applied to every point in the output image grid (by using values from a DEM), prior to resampling. As this is a computationally expensive exercise it may be beneficial to perform some interpolation between points on a 'super-grid', particularly where relief does not vary greatly over large horizontal distances, or in nadir areas of an image where the effect would not be expected to be pronounced.

Because of the stability of the SPOT geocoding model it can be applied to several adjacent scenes using a technique known as pass processing. This involves taking the GCP adjustment applied to a single scene and applying it to a number (typically six to ten) of images in a single pass (Swann *et al.* 1988). This is useful for ocean areas or featureless interiors such as Australia or the Antarctic where good ground control is difficult and expensive to establish. Using this method, average requirements of less than one GCP per image can be attained.

1.5 AVHRR GEOCODING MODELS

1.5.1 Bulk AVHRR geocoding

The AVHRR is one of the primary instruments carried aboard the NOAA series of polar-orbiting satellites and provides daily coverage of the entire globe at a resolution of 1 km. A summary of the AVHRR technical specification can be seen in table 1.7.

Table 1.7. AVHRR technical specification

Country of orgin	United States
Launch	NOAA-6—27 June 1979 (followed by NOAA-7 to NOAA-13)
Orbit	sun-synchronous (near polar)
Orbit period	102 minutes
Inclination	98.7° to 98.9°
Altitude	833–870 km
Coverage	Global
Repeat cycle	12 hours
Swath width	2700 km
Pixel size	1.1 km (nadir)
Band 1	0.58–0.68 μm
Band 2	0.72–1.10 μm
Band 3	3.55–3.93 μm
Band 4	10.3–11.3 μm
Band 5	11.3–12.5 μm

Although originally designed as a meteorological satellite AVHRR has found a vast number of other applications, due to its synoptic coverage and rapid revisit capability (see for example section 3.3.3 or section 5.5.3). The TIROS-N satellites carrying the sensor are in a sun-synchronous orbit at an altitude of 833–870 km. The repeat cycle of twelve hours is significantly different from that of Landsat or SPOT. The satellite does not have sufficient onboard recording capacity to store an entire global coverage of 1 km data known as local area coverage (LAC), so for fully global cover it is resampled on board (Kidwell 1986). The resampling is performed by averaging the first four pixels out of every five and leaving a gap of 3 km across scan-lines; this is often regarded in practice as being equivalent to producing imagery of 4 km resolution. As with high-resolution sensors, geocoding of AVHRR is an essential prerequisite to accurate information extraction. There are, however, major differences in the geometry of the images being considered, for example:

(1) the area being imaged means that there is significant distortion induced by the curvature of the earth;
(2) the large pixel size means that it is not necessary to transform the images to the same absolute geometrical accuracy;
(3) large-scale images are not susceptible to the same local relief effects that Landsat and SPOT images suffer from.

The earth curvature effect is caused by the large swath width and the fact that the AVHRR pixels each represent a fixed angle rather than a fixed size on the ground. Consequently pixels near the edge of an image are much larger than those at the centre. The relationship between the geocentric angle and the scanning radiometer angle (Brush 1988) can be seen in figure 1.8.

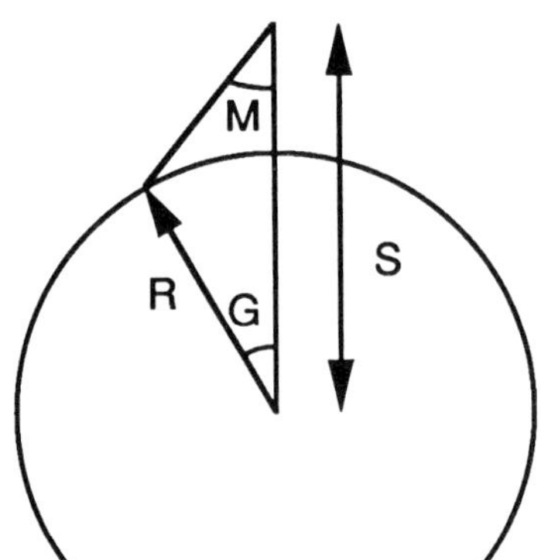

Fig. 1.8. Derivation of the geocentric angle for the AVHRR sensor (adapted from Brush 1988).

Using the trigonometric sine rule yields the following formula:

$$\frac{\sin(G+M)}{S}=\frac{\sin(M)}{R} \tag{1.11}$$

where

$G =$ geocentric angle from nadir,

$M =$ scanning radiometer scan angle,

$S =$ satellite distance from the geocentre,

$R =$ earth's radius.

From this it is possible to convert between the two angles and to resample to a constant geocentric angle, hence a constant pixel size.

1.5.2 Precision model

More accurate corrections for image distortion can be obtained by considering a simplified model of the satellite orbit (Emery *et al.* 1989). This models the satellite position and sensor-pointing direction based on the readily available satellite ephemeris data. The model used must also be able to incorporate the effect of earth rotation and sensor mirror dynamics. The scanner optics consist of a telescope fixed relative to the spacecraft and a mirror at 45° to the telescope. The rotation of the mirror directs radiation reflected from the earth's surface (or from clouds) into the telescope optics. The mirror rotates at a constant rate (6 times per second) relying on the satellite's forward motion to allow a scanned image to be built up. The simple scanner optics make it relatively straightforward to formulate a mathematical sensor model. Because of the large pixel size it is rarely necessary to compensate for small-scale terrain distortions, thus a model of the earth's surface as an ellipsoid (or even a sphere) will generally suffice. The remaining geometric distortions can then be eliminated by a standard template matching technique using GCPs. For example, timing inaccuracies will often lead to a transformation which is intrinsically correct but displaced along the track. As a rule of thumb 6 to 10 GCPs distributed evenly in the north–south direction will normally suffice to perform the necessary along-track nudge, and even a single GCP (Ho and Asem 1986) can be sufficient to produce an RMS error of better than 2 pixels (approximately 2 km). Alternatively, because of the large area involved, it may be preferable to allow an operator to navigate the image interactively using a digitised coastline.

It is also important to be able to transform large area images into a suitable map projection, for example AVHRR products available from NOAA are provided in three standard projections namely polar stereographic, plate Carree and Mercator. Each of these has certain disadvantages for large area applications, for example the Mercator introduces significant distortion in polar regions and the polar stereographic splits the earth into two distinct hemispheres. Another disadvantage of these particular projections is that they are not 'equal area' in that the true area represented by an individual pixel is not a constant but is a function of latitude. For this reason it may be desirable to transform AVHRR data into an equal-area radial projection such as Peters' projection (Lloyd and d'Souza 1987).

1.5.3 Automatic methods

As with high-resolution satellite imagery it is beneficial to automate the geocoding of AVHRR images by using pattern matching techniques to relocate GCPs (Sun and Takagi 1987). The basic method is to produce a library of GCP chips which can be used to

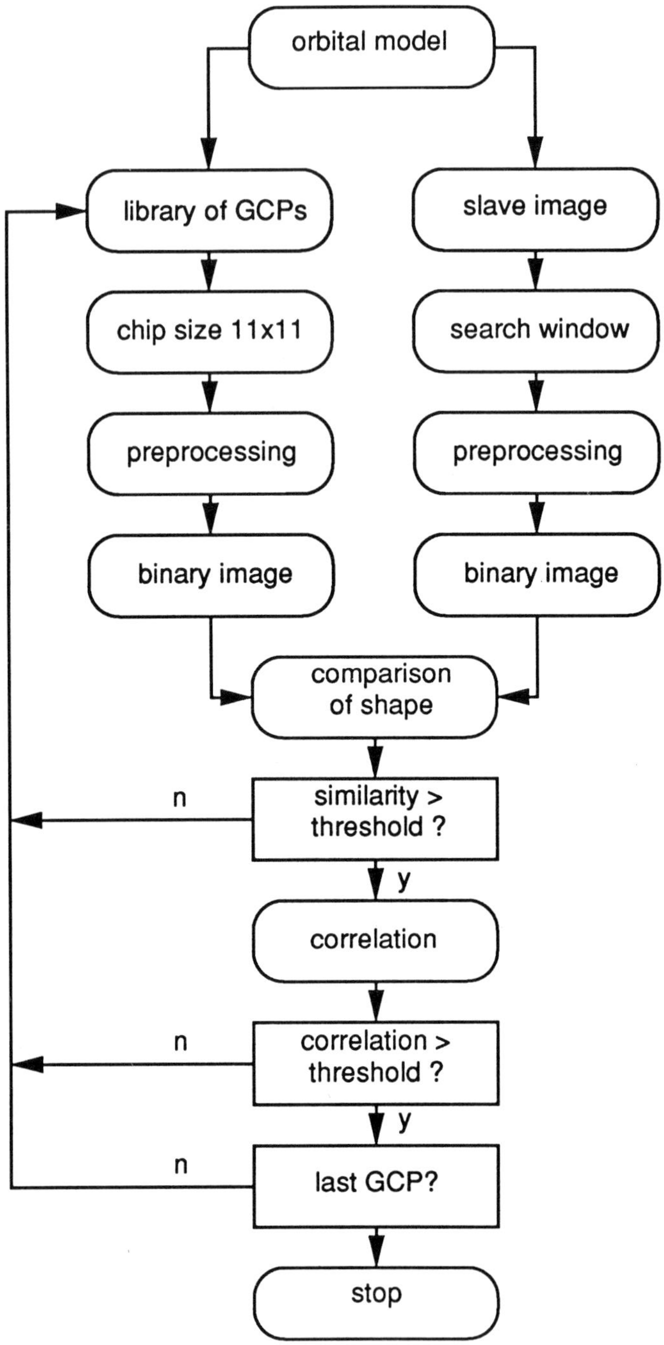

Fig. 1.9. Operational flowlines for AVHRR geocoding using image chips (adapted from Cracknell and Paithoonwattanakij 1989).

automatically determine the location of control points, for example (Cracknell and Paithoonwattanakij 1989) Dundee University have built a library of 337 AVHRR ground control points of the UK. This is particularly valuable as multiple AVHRR images of the same area are acquired frequently (sometimes several a day) and it is imperative to geocode them in near real time. The searching algorithm is seeded by providing approximate locations of three GCPs; because the greatest distortion in AVHRR imagery is in the extreme off-nadir pixels it is advantageous to provide GCPs in the form of a triangle with vertices at (top line, centre), (bottom line, extreme left), (bottom line, extreme right). Alternatively the AVHRR orbit model can be used to determine approximate seed point locations automatically. A typical flowline for automatic geocoding of AVHRR imagery can be seen in figure 1.9.

More accurate chip relocation can be attained by using high-resolution chips matched to the low-resolution AVHRR images, for example (Cracknell and Paithoonwattanakij 1989) Landsat-MSS data can be used to achieve registration accuracy of better than 0.25 pixel (approximately 250 m). Sub-pixel geocoding is necessary for a number of applications, for example derivation of ocean current vectors. With conventional geocoding accuracy (of the order of a pixel at nadir) the current vectors can be measured to an accuracy of ± 1.8 cm s^{-1}; if a geocoding accuracy of 0.2 km is available then they may be measured to an accuracy of ± 0.5 cm s^{-1}.

1.6 REFERENCES

Adamson J, Kerr G W and Jacobs G H P (1988) Rectification quality assessment of Meteosat images. *ESA Journal*, **12**, 467–482.

Benny A H (1983) Automatic relocation of ground control points in Landsat images. *International Journal of Remote Sensing*, **4** (2) 335–342.

Benny A H (1985) Example of the use of Landsat satellite imagery for the accurate location of offshore islands. *International Journal of Remote Sensing*, **6** (10) 1581–1584.

Binnie D R and Colvocoresses A P (1987) The Denali image map. *Photogrammetric Engineering and Remote Sensing*, **53** (3) 307–310.

Brooks J, Kumar R and Levine I (1981) Implementation of the Space Oblique Mercator projection in a production environment. *Proceedings of the Fifteenth International Symposium on Remote Sensing of Environment*, Ann Arbor, Michigan, May 1981, 671–674.

Brush R J H (1988) The navigation of AVHRR imagery. *International Journal of Remote Sensing*, **9** (9) 1491–1502.

Clark B P (1990) Landsat Thematic Mapper data production—a history of bulk image processing, *Photogrammetric Engineering and Remote Sensing*, **56** (4) 447–451.

Cracknell A P and Paithoonwattanakij (1989) Pixel and sub-pixel accuracy in geometrical correction of AVHRR imagery *International Journal of Remote Sensing*, **10** (4 and 5) 661–667.

Curlander J C, Kwok R, Pang S S and Pang A A (1987) An algorithm for generation of geocoded data products from spaceborne SAR imagery. Jet Propulsion Laboratory,

California Institute of Technology, Pasadena, California, JPL D-4801, September 21, 1987.

Davison G J (1986) Ground control pointing and geometric transformation of satellite imagery. *International Journal of Remote Sensing*, **7** (1) 65–74.

Derenyi E and Pollock R (1990) Extending a GIS to support image-based map revision. *Photogrammetric Engineering and Remote Sensing*, **56** (11) 1493–1496.

Emery W J, Brown J and Nowak Z P (1989) AVHRR image navigation: summary and review. *Photogrammetric Engineering and Remote Sensing*, **55** (8) 1175–1183.

Ford G E and Zanelli C I (1985) Analysis and quantification of errors in the geometric correction of satellite images. *Photogrammetric Engineering and Remote Sensing*, **51** (11) 1725–1734.

Friedmann D E, Friedel J P, Magnussen K L, Kwok R and Richardson S (1983) Multiple scene precision rectification of spaceborne imagery with very few control points. *Photogrammetric Engineering and Remote Sensing*, **49** (12) 1657–1667.

Harrison B A, Jupp D L B, Hutton P G and Mayo K K (1989) Assessing remote sensing technology, the microBRIAN example. *International Journal of Remote Sensing*, **10** (2) 301–309.

Heard M I, Mather P M and Higgins C (1992) GERES: a prototype expert system for the geometric rectification of remotely-sensed images. *International Journal of Remote Sensing*, **13** (17) 3381–3385.

Ho D and Asem A (1986) NOAA AVHRR image referencing. *International Journal of Remote Sensing*, **7**, 895–904.

Hough H (1992) Satellite GPS and remote sensing. *GPS World*, February 1992, 18–24.

Kidwell K B (1986) *NOAA Polar Orbiter Data User's Guide, NOAA NESDIS*. National Climate Data Centre, Washington DC, USA.

Lloyd D and d'Souza G (1987) Mapping NOAA-AVHRR imagery using equal-area radial projections. *International Journal of Remote Sensing*, **8** (12) 1869–1878.

Mack G A H (1992) Skyfix—a better position through the use of Inmarsat and GPS. *Proceedings of Satellite Symposia 1 and 2 : Navigation and Mobile Communications, and Image Processing, GIS and Space-assisted Mapping, from the 'International Space Year' Conference*, Munich, Germany, 30 March–4 April 1992, (ESA ISY-2), 81–86.

Moik J G (1980) *Digital Processing of Remotely Sensed Images*. National Aeronautics and Space Administration, Washington DC, NASA SP-431.

O'Brien D M and Prata A J (1990) Navigation of ERS-1 Along Track Scanning Radiometer (ATSR) images. *ESA Journal* **14** (4) 447–465.

Sharun C (1993) GPS and GIS: operations use of GPS for cutblock layout under forest canopy. GIS '93 Symposium, Vancouver, British Columbia, February 1993, 189–193.

Shlien S (1979) Geometric correction, registration and resampling of Landsat imagery. *Canadian Journal of Remote Sensing*, **5**, 74–87.

Sun W and Takagi M (1987) Geometric distortion correction with high accuracy for NOAA satellite images. *Proceedings of the 1987 International Geoscience and Remote Sensing Symposium*, Ann Arbor, Michigan, 1257–1262.

Swann R, Hawkins D, Westwell-Roper A and Johnstone W (1988) The potential for automated mapping from geocoded digital image data. *Photogrammetric Engineering and Remote Sensing*, **54** (2) 187–193.

Westin T (1990) Precision rectification of SPOT imagery. *Photogrammetric Engineering and Remote Sensing*, **56** (2) 247–253.

Wilson A K (1988) The effective resolution element of Landsat Thematic Mapper. *International Journal of Remote Sensing*, **9** (9) 1303–1314.

2

Synthetic aperture radar geocoding

Chapter 2 describes techniques and algorithms used for the geocoding of synthetic aperture radar (SAR) imagery which is becoming an increasingly important source of geographic information. It contains the following sections:

— the SAR imaging mechanism: how SAR images are produced and their fundamental difference from optical images;
— geometric aspects of SAR: the fundamental geometry of SAR images and characteristic phenomena such as shadowing, foreshortening and layover;
— bulk geocoding: the SAR geometry equations (doppler shift, range, and earth model) and how they are solved to produce a bulk geocoded image;
— precision geocoding: the use of DEMs to correct for terrain effects, ground control pointing and validation of the geocoded product;
— combination with optical data: the benefits and methods needed to combine geocoded SAR with other images.

2.1 THE SAR IMAGING MECHANISM

All the major satellite sensors that have been described so far (such as Landsat, SPOT and AVHRR) have certain characteristics in common. The most important of these is that they are passive sensors relying for their energy source on either reflected sunlight, or in the case of thermal sensors, electromagnetic radiation from warm bodies. One consequence of this is that the sensors can only operate in relatively narrow spectral windows due to the atmospheric attenuation of incoming and reflected radiation at most spectral frequencies. Another, more serious, limitation is the inability of the sensors to operate in cloudy or hazy conditions. This can seriously impede operations in temperate regions such as northern Europe or in tropical regions when the cloudy seasons are often of the most interest. This has led to an increased utilisation of active microwave sensors which generate their own pulse of energy and then record the signal obtained by its reflection from the earth's surface.

One of the most important spaceborne microwave sensors is the synthetic aperture radar (SAR), examples of which were flown on the US Seasat mission in 1978, on two

NASA space shuttle missions (as the Shuttle Imaging Radar, SIR-A and SIR-B) and as the primary instrument on the European Space Agency ERS-1 mission, the Japanese counterpart (JERS-1) and the planned Canadian Radarsat mission (see figures 2.1 and 2.2 in the section between pp. 84 and 85).

A summary of the major spaceborne SAR missions can be seen in table 2.1 with more detailed technical specifications in section 5.3.

Table 2.1. Major spaceborne SAR missions

Mission	Country	Launch
Seasat	USA	1978
SIR-A	USA	1981
SIR-B	USA	1984
ERS-1	Europe	1991
JERS-1	Japan	1991
ERS-2	Europe	(due 1995)
SIR-C	USA	1994
Radarsat	Canada	(due 1995)

Although images produced from SAR appear similar to those obtained from optical/infrared imagery they have several significant differences. For a start, because SAR sensors are not reliant on the sun's energy for illumination their operation is far less affected by the time of year, the season of acquisition and the prevalent weather. The imaging mechanism is also fundamentally different—an optical sensor forms an image whose features are determined by the angles subtended at the sensor by objects on the earth's surface whereas the basic operation of a radar system is to time the return of an electromagnetic wave from when it leaves a transmitting antenna to when it returns to the receiving antenna. Because the speed of the electromagnetic waves is a constant (the speed of light) there is a direct relationship between the time taken for the signal to return and the distance of the target being detected. The two main factors that determine the nature of the reflected radar beam are:

(1) the parameters of the imaging system (such as incidence angle, resolution, wavelength and polarisation);
(2) the properties of the target being imaged (such as the surface roughness, the dielectric constant, moisture content of the target, and local topography).

A radar that can be used to form an image of the earth's surface is known as an imaging radar sensor. Non-imaging sensors, such as the scatterometer and the radar altimeter are also widely used for earth observation. Imaging radars are classified into two types:

— real aperture radars (more usually flown on aircraft);
— synthetic aperture radars (flown on both satellite and aircraft platforms).

The basic concept behind a SAR is to make use of the satellite motion to synthesise a larger antenna from the position of the platform at a sequence of points. The return from a point on the earth's surface is detected at a number of different positions and the signals are corrected for phase and combined by a method known as coherent addition. The image format as generated by a SAR system can be seen in figure 2.3.

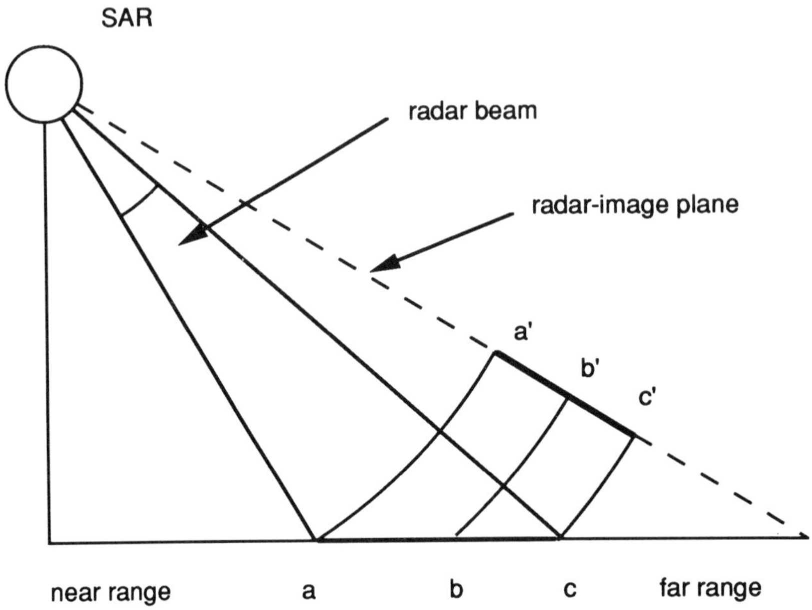

Fig. 2.3. Basic SAR image format on a level surface - the relationship between ground range images and slant range images can be seen by comparing ground points (marked *a b c*) with their counterparts in the radar image plane *a′ b′ c′* (adapted from Keydel 1992).

SAR data is amenable to precise geolocation (Schreier 1991) because time measurements in SAR are more precise than angular measurements in optical imagery. Given a sufficiently accurate orbital model it should therefore be possible to produce a precision geocoded product.

2.2 GEOMETRIC ASPECTS OF SAR

The geometry of SAR images has several differences from that of optical imagery. Two of the geometric aspects of SAR which are not found in optical images are shadowing and layover. In an optical image the side of a raised feature such as a hill opposite to the illumination source will not be imaged if it is in an area of shadow. Radar images will also have the same areas of shadow, however as long as there are no other surface targets at the same distance from the sensor the shadowing effect will cause the appearance of darker 'missing' features in the image. The layover effect is caused by objects at the same distance being imaged together. This is an example of the more general problem of ambiguous imagery whereby two distinct objects from different surface features are mapped onto the same resolution cell (pixel). The counterpart in optical images occurs when objects at the same angular position (that is on the same optical ray) but at different distances are imaged simultaneously—in this case the ambiguity is resolved because only the front object is actually imaged. These effects are illustrated in figure 2.4. In both types of imagery the effects will be most apparent in areas of high relief, particularly where the local terrain slope is similar or greater than the incidence angle of the sensor.

In a similar manner to optical images, SAR images also experience a relief effect due to local topography. On a slope facing toward the sensor neighbouring pixels will be roughly the same distance from the sensor resulting in a foreshortening effect. In the extreme case when the terrain slope is greater than that of the incident signal layover will occur. On slopes facing away from the sensor the features will be portrayed more accurately, although in the limiting case there may be a shadowing effect. These effects not only produce geometric distortion but also disturb and destroy the backscatter information—geometric and radiometric relations for SAR are even more intricately intertwined than with optical images, for example the images of Mount Shasta in California, shown in figure 2.5 in the plate section between pp. 84 and 85, demonstrate the effect on image appearance of two different viewing angles.

2.3 BULK GEOCODING

For both optical and SAR imagery the basic purpose of geocoding is to associate an accurate terrestrial location with every point on an image—however with optical images

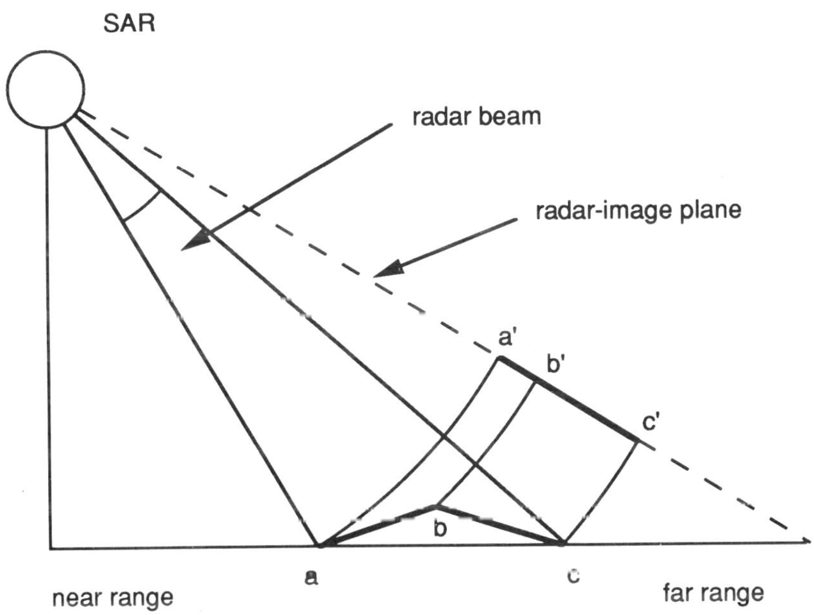

Fig. 2.4(a)

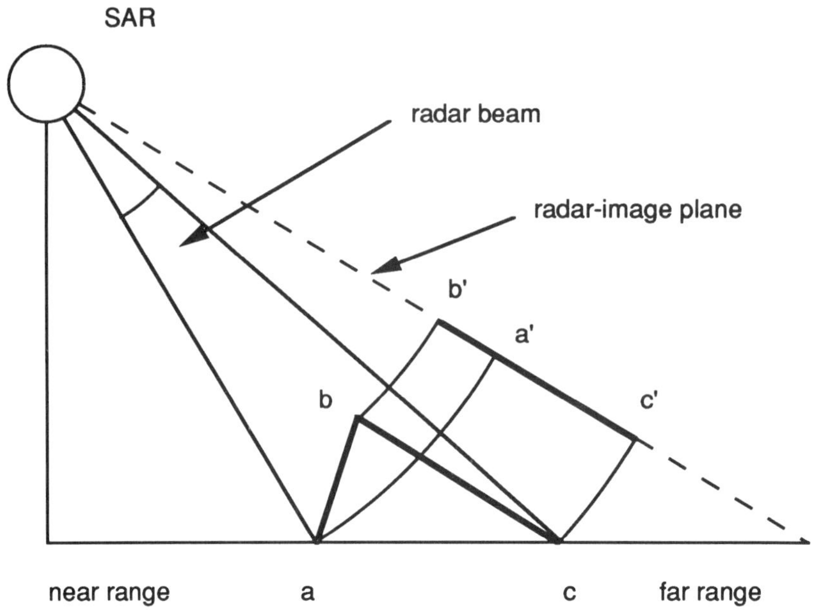

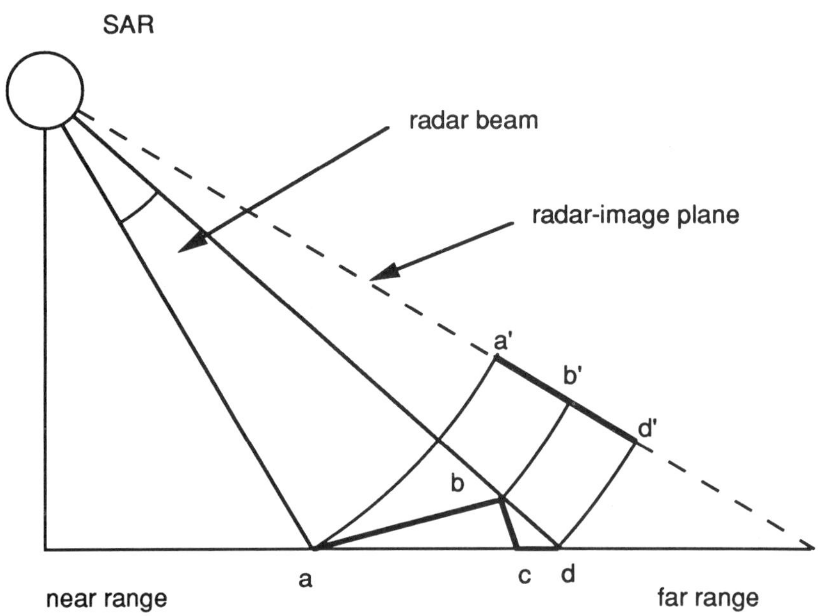

Fig. 2.4. Geometric effects in SAR images: (a) foreshortening—slope *ab* is inclined toward the radar and slope *bc* is inclined away from the radar (back slope). The distance corresponding to *ab* in the image plane is much shorter than that corresponding to slope *bc*; (b) layover—the angle of the slope is greater than the look-angle, the slant range for point *a* is now greater than that for point *b* and their relative positions (*a′b′*) have been reversed in the image plane; (c) shadow—the slope *bc* causes a radar shadow in the image plane (adapted from Keydel 1992).

it is the spacecraft position and angular measurements that are the governing factors; for SAR it is radar time delays and shifts in signal frequency. The basic task of SAR geocoding is to relate the time and range information obtainable from the radar to the corresponding location on the earth's surface. The three fundamental equations which determine the SAR geometry (Curlander *et al.* 1987) are:

(1) the doppler shift equation;
(2) the range equation;
(3) the earth model equation.

By solving all three equations simultaneously the geodetic location of the required point (x, y, z) can be found.

The doppler phenomenon describes the way that, within the radar footprint, signals reflected from targets moving towards the sensor are shifted to higher frequencies and those moving away from the target are shifted to lower frequencies. The digital processor filters the signal to focus the energy towards one specific point in the azimuth line. Most processors utilise the point of no frequency shift (zero doppler) as reference in the azimuth focusing. For a point target the doppler shift can be estimated from the formula:

$$f_\mathrm{d} = \left(\frac{2}{\lambda R}\right)(\boldsymbol{V}_\mathrm{s} - \boldsymbol{V}_\mathrm{t})\cdot(\boldsymbol{R}_\mathrm{s} - \boldsymbol{R}_\mathrm{t}) \tag{2.1}$$

where

f_d = the doppler value used in the azimuth phase reference relation,

λ = radar wavelength,

R = sensor–target slant range,

$\boldsymbol{V}_\mathrm{s}$ = sensor velocity vector,

$\boldsymbol{V}_\mathrm{t}$ = target velocity,

$\boldsymbol{R}_\mathrm{s}$ = sensor position vector,

$\boldsymbol{R}_\mathrm{t}$ = target position vector.

Note that to use this equation it is necessary to have imagery in slant range format—this is the format generated by most processors (Schreier 1991). In slant range imagery one axis represents the distance from the satellite position to the target, whilst the orthogonal axis represents the doppler shift.

In general targets such as fields and roads will be stationary relative to the earth's surface, thus

$$\boldsymbol{V}_\mathrm{t} = \boldsymbol{\omega}_\mathrm{E} \times \boldsymbol{R}_\mathrm{t} \tag{2.2}$$

where

$\boldsymbol{V}_\mathrm{t}$ = target velocity,

$\boldsymbol{\omega}_\mathrm{E}$ = earth's rotational velocity (7.292×10^{-5} rad/sec),

$\boldsymbol{R}_\mathrm{t}$ = target position vector.

From simple geometry the slant range from a sensor to the target is given by

$$R(i,j) = \{(\boldsymbol{R}_s - \boldsymbol{R}_t) \cdot (\boldsymbol{R}_s - \boldsymbol{R}_t)\}^{1/2} \tag{2.3}$$

where

$R(i,j) =$ slant range at pixel (i,j),

$\boldsymbol{R}_s =$ sensor position vector,

$\boldsymbol{R}_t =$ target position vector.

Finally, as with optical images a suitable basic model for the earth can be that of a sphere or an ellipsoid (see section 1.3.3).

2.3.2 Geocoding process

In simple terms the geocoding process will consist of the following phases:

— calculate the components of the model;
— calculate the transformation and apply it to points on the output grid;
— resample using the original data.

The initial estimate of the imaging model requires a spacecraft orbit model which is based on the transmitted satellite ephemeris. Following this, various algorithms can be used to determine the geographic location of a pixel (x, y). As with the geocoding procedure described in chapter 1 the basic requirement (Coll and Pettigrew 1988) is to be able to determine a transformation between a regular grid in the desired output space and the uncorrected SAR image. In practice this reverse transformation is not simple to evaluate directly so the approach usually adopted is to determine the corresponding forward transform (from the uncorrected imagery to the map space) and then to invert it numerically. This process is best performed iteratively, for example (Kwok and Curlander 1987):

(1) determine the nominal target location by intersecting the range vector with the geoid;
(2) calculate the target doppler vector;
(3) calculate iso-doppler lines (points of the earth's surface with constant doppler frequency f_d);
(4) maintaining the correct sensor–target range rotate the range vector to intersect the appropriate iso-doppler line given by f_d;
(5) iterate this process until f_d converges to a predetermined value.

As with optical imagery, the reverse transformation is used to map back from the desired output grid to the raw input data. In order to preserve information it is highly desirable to only use one resampling step—this is another reason to use slant range data rather than that which has already been resampled into ground range. Interpolation kernels are generally similar to those used for optical imagery (for example nearest neighbour, bilinear interpolation and cubic convolution). More sophisticated, higher-order kernels such as the discrete fourier transform (DFT) have also been proposed as suitable for SAR geocoding (Laycock 1990). The DFT is another approximation to the sinc interpolator and, as such, suffers from a general problem associated with truncated interpolators, that it is possible to generate negative image values. By setting these values to zero the overall image statistics become biased; however it appears that this problem is less severe when using multilook SAR images than it is with optical images.

2.4 PRECISION GEOCODING

2.4.1 Terrain correction

Because of the nature of SAR geometry, even in areas of moderate changes in relief the use of a DEM for geocoding becomes essential. Various alternative approaches, using DEMs to perform range displacement correction are possible. For example, figure 2.6 shows the data processing flow for a three-pass correction scheme.

The functions performed in each pass are as follows:

First pass

— rectify slant range image and map in the azimuth direction.

Second pass

— generate distortion map to account for relief displacements;
— combine with resampled locations from first pass;
— resample first pass image in range.

Third pass

— complete resampling to geocoded map format.

For operational SAR geocoding it is advantageous to maintain databases of DEMs and GCPs, for example the SAR geocoder (Schreier 1991) installed at the DLR processing and archiving facility for ERS-1 uses three major databases:

(1) a map library database storing the features of 8000 topographic maps of areas in Europe;
(2) a GCP database, including chips extracted from SAR images;
(3) a DEM database where the data is stored in a common geographic format but with applications modules that permit the DEMs to be extracted in any recognised European map projection and at a suitable grid spacing.

DEMs can also be used (Schreier *et al.* 1990), in an alternative method of precision geocoding, to generate a synthetic image—the appearance of a SAR image is dictated to a far greater extent than optical imagery by the terrain, making it possible to simulate the appearance of a SAR image based on a DEM and known radar parameters such as incidence angle. This can then be used for precise geometric control by registering an image to the synthetic image. This can usually be achieved by simple correlation techniques and has the advantage of producing increased geometrically fidelity, since the data set being used for ground control is the same as that used for terrain correction. Another advantage is that it is possible to locate areas of difficult topography in advance and to

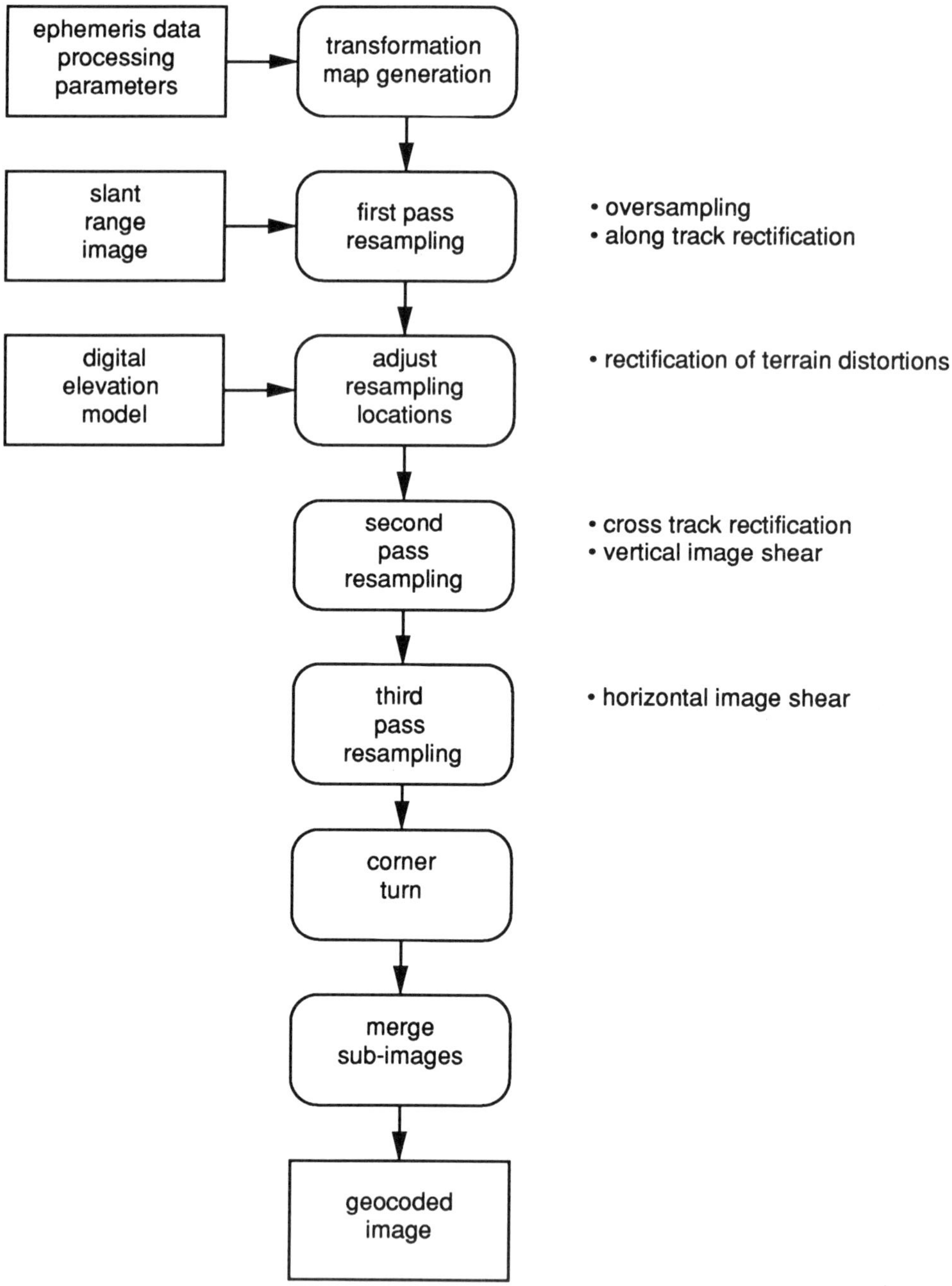

Fig. 2.6. Dataflow for SAR geocoding using a three-pass resampling method (adapted from Kwok and Curlander 1987)

pre-calculate simulated sub-images for these areas. Another use of simulated images (Tilley and Bonwit 1989) is to reduce layover distortion in images. The effect of layover is to concentrate the radar returns from slopes that face towards the radar receiver. The simulated image will show where this effect has taken place and a suitable redistribution of energy 'up slope' can be performed. One final use of DEMs in SAR geocoding is to 'pre-screen' hilly areas where terrain correction will be necessary.

2.4.2 Use of ground control points

Even with the use of relatively accurate DEMs there is still likely to be residual low-frequency errors which are most easily removed, as in the case of optical images, by the use of GCPs. The main sources of these residual errors (Schreier *et al.* 1990) are:

(1) remaining uncertainties in pixel location (estimated at 20 metres in range and 110 metres in azimuth in the case of ERS-1);
(2) shifts and scalings relative to the common ellipsoid (possibly up to 300 metres);
(3) errors in DEM generation (for example the relatively accurate DEMs available for Germany are estimated to have errors of approximately 30 metres in plan and 7 metres in elevation).

The desirable attributes for GCPs for use in SAR geocoding (Weiler *et al.* 1986) are similar to the optical case and include:

(1) they are readily detectable and distinct from the surrounding area;
(2) they are well distributed throughout the image;
(3) they can be locatable on suitable maps (or their position accurately known for example by GPS survey);
(4) they should remain stable over time.

Using GCPs as part of a precision geocoding model it is possible to achieve sub-pixel geocoding accuracy even in areas of rough terrain. For example, geocoded Seasat SAR imagery of Canada with a nominal resolution of 25 metres (Coll and Pettigrew 1988) achieved an RMS accuracy of 12.9 metres (after errors due to map measurement had been subtracted) on an image of moderate terrain relief and a similar result on an area with noticeable terrain relief.

The automatic selection of GCPs from SAR images is more problematical than with optical images because of their radiometric characteristics, primarily speckle—although this can be reduced to a certain extent by the use of non-linear filters such as the Lee sigma filter (Lee 1981). Even following such an operation correlation methods to relocate control points may not prove successful. One alternative approach (Weiler *et al.* 1986) is the use of an 'interest operator' which identifies points in an image which are clearly distinguishable from the background. The algorithm used is as follows:

(1) for each block of pixels in the image calculate a threshold;
(2) for each 4×4 window in the block calculate the variance of the pixel values in window;
(3) if (variance > threshold) then set output pixel to 1, else set output pixel to 0.

The appropriate threshold is calculated by setting up a grid of regularly spaced 4×4 windows. For each window the variance of the pixel values in the window is calculated. The variances are then sorted into descending order and a weighted mean calculated using the vector $\{0, 15, 10, 6, 3, 1, 0, 0, 0, 0, 0, 0, 0, 0, 0, 0\}$.

Finally if several contiguous images from the same orbit are required for geocoding, a pass-processing scheme (Coll and Pettigrew 1988) analogous to that for optical images described in section 1.4 may be employed.

2.4.3 Validation

In the case of SAR geocoding the 'impossible' points that blunder detection needs to avoid are those where there is no information to reconstruct the geocoded image (shadow areas) or where there is insufficient information to reconstruct the original signal accurately (layover areas); these areas are then omitted from the final geocoded product. It is often desirable to derive this information as a parallel process of the SAR geocoding, for example the SAR geocoder (Schreier 1991) installed at the DLR processing and archiving facility produces an accompanying 'layover, shadow and incidence angle mask' which indicates the incidence angle and areas affected by layover and shadow. Blunder detection can also be applied to filter out 'bad' GCPs in the standard manner, for example the following rules (Raggam 1990) can be used to remove blunders in GCP selection:

— eliminate GCPs with residuals greater than three times the RMS error;
— only use the best *n* GCPs;
— use a predefined residual limit and constrain the distribution of GCPs to be homogeneous over the image area.

The validation of the geocoding process itself (Dowman and Upton 1990) includes statistical and radiometric measures of quality and evaluation of error equations predicted from sensor and terrain parameters. Another set of validation parameters is available in the form of residual errors if GCPs are used. Product validation includes comparison of the output product with reference maps and comparison with synthetic images which use slope values derived from DEMs to predict the expected backscatter. A summary of validation functions can be seen in table 2.2.

Table 2.2. Validation functions for SAR geocoding (based on Dowman and Upton 1990)

Validation function	Quality metric
Compare layover and shadow with geocoded product	shadow/layover masks histograms
Evaluate error equations	predicted errors from geocoding model
Calculate statistical quality measures	radiometric quality
Compare geocoded product with maps/DEMs	residual vectors profile statistics

2.5 COMBINATION WITH OPTICAL DATA

Registration of SAR images with geocoded optical data can be useful for a variety of applications, from the generation of combined geocoded products to data fusion for enhanced information extraction. Two basic methods which can be used for multisensor registration with SAR are matching of scene features and using a DEM to generate a synthetic SAR image, to which the actual image can be registered. In a typical approach to feature-based registration (Rignot *et al.* 1991) the objective is to commence with a set of images that have been precision geocoded to within 2 pixels and then to fine-tune the registration using suitable methods (see figure 2.7).

This approach is particularly suitable where there is little ancillary data in the form of maps, digital elevation models or ground control points available. The first step is to automatically select a series of sub-frames from the image where registration can be performed with confidence. This requires the identification of a robust set of invariant features which can be unambiguously identified in all the images to be registered. They must, therefore, be visible at a variety of resolutions and imaging conditions (such as view angle, presence of speckle) and be relatively stable across time. Candidate features include roads, coastal features and seasonally stable water features (such as reservoirs). Features can be identified interactively using an IIPF or automatically by constructing an edge map and using a figure of merit approach (Davis and Kenue 1978) to locate the best of the features. The registration stage then consists of three operations:

— feature extraction;
— feature matching;
— constraint filtering.

Feature extraction

Feature extraction can be performed using a variety of techniques (some of which are described in more detail in section 4.2.2). Two such techniques are region boundary extraction and edge detection. Region boundary extraction uses the fact that although local regional brightness values may vary greatly at the same location in a multisensor dataset, regions with homogeneous statistical properties (such as mean grey-level and texture) that correspond to real world features should occur in the same location across the various datasets. Because of the speckle present in SAR images, methods of segmentation based on grey-level alone tend not to be successful and segmentation is usually performed using a combination of local mean grey-level and local texture values (Rignot *et al.* 1991). The edge detection approach can also be used to extract features, but again account must be taken of the speckle present in SAR imagery. One approach is to use a gaussian smoothing operator which approximates to an optimal operator (detects edges well, locates their positions accurately, does not provide a multiple response to a single edge) such as a Canny operator (Canny 1986) or the Marr-Hildreth operator (Marr and Hildreth 1980).

Feature matching

The output from the feature extraction stage is a series of binary feature maps for each input image simply indicating the presence (or absence) of a feature at a particular

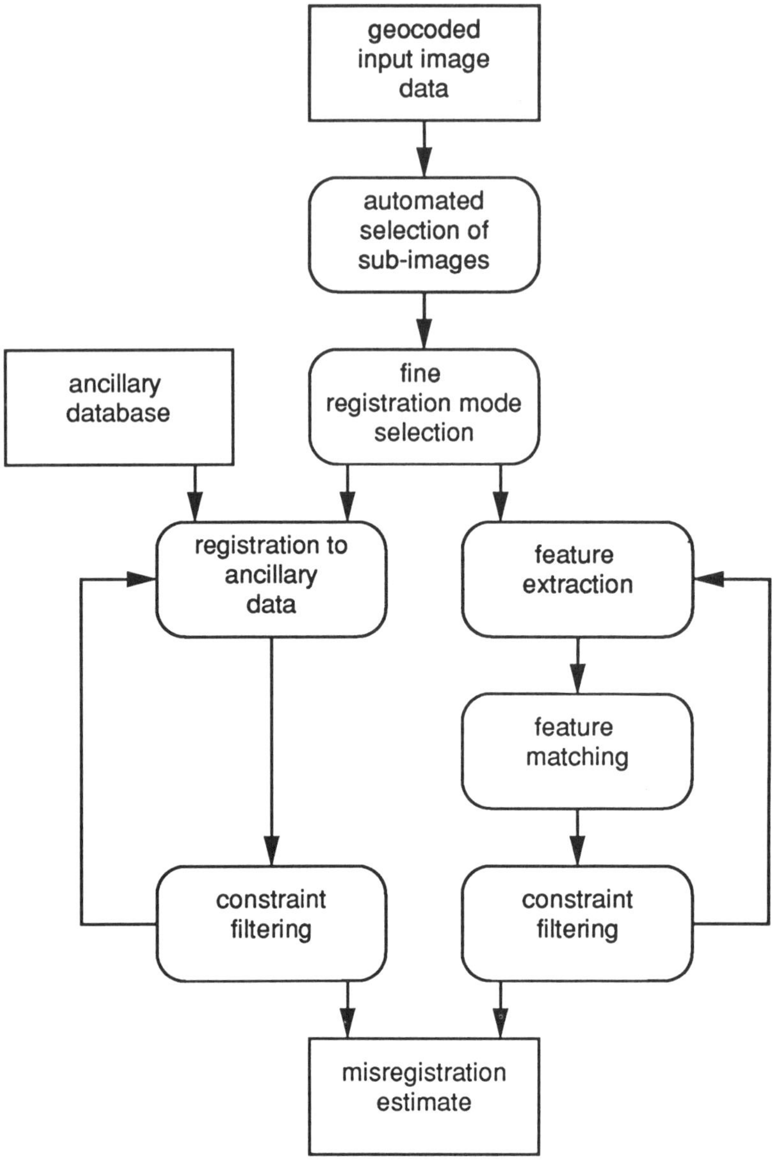

Fig. 2.7. Flowlines for a multisensor registration algorithm (based on Rignot *et al.* 1991).

location. The simplest method for finding the optimum match between the feature maps is binary correlation; other methods include the distance transform (Barrow *et al.* 1977) where the binary edge maps are transformed into grey-level distance maps by allocating to each pixel the distance to the nearest feature (represented by the nearest point with a value 1 in the feature map). The relative shifts between the image locations that produces the global minimum in differences between the distance maps is used as the best-fit registration.

Constraint filtering
Finally, constraint filtering (comparing the results obtained from several independent estimates of registration accuracy) can be used to validate the results of the feature matching. If the estimates are inconsistent the feature extraction and feature matching stages are repeated until a consistent result is obtained.

2.6 REFERENCES

Barrow H G, Tenenbaum J M, Bolles R and Wolf H C (1977) Parametric correspondence and chamfer matching: two new techniques for image matching. *Proceedings of the Fifth Joint Conference on Artificial Intelligence*, Cambridge, Massachussets, 1977, 659–663.

Canny J F (1986) A computational approach to edge detection. *IEEE Transactions on Pattern Analysis and Machine Intelligence*, **8** (6) 679–698.

Coll E C and Pettigrew R G (1988) A high fidelity, high throughput system for geocoding SAR imagery. *Proceedings of IGARSS '88 Symposium*, Edinburgh, Scotland, 13–16 September 1988 (ESA SP-284), 687–690.

Curlander J C, Kwok R, Pang S S and Pang A A (1987) *An Algorithm for Generation of Geocoded Data Products from Spaceborne SAR Imagery*. Jet Propulsion Laboratory, California Institute of Technology, Pasadena, California, JPL D-4801, September 21, 1987.

Davis W A and Kenue S K (1978) Automatic selection of control points for the registration of digital images. *Proceedings of the Fourth International Conference on Pattern Recognition*, Kyoto, Japan, 7–10 November 1978, 936–938.

Dowman I and Upton M (1990) A Sun environment for SAR geocoding. *Proceedings of the Third International Workshop on Image Rectification of Spaceborne Synthetic Aperture Radar*, Farnham Castle, UK, 1–3 October 1990, 81–85.

Keydel W (1992) Basic principles of SAR, in *Fundamentals and Special Problems of Synthetic Aperture Radar (SAR)*. Advisory Group for Aerospace Research and Development (AGARD), AGARD-LS-182, August 1992, 1–13.

Kwok R and Curlander J C (1987) Automated rectification and geocoding of SAR imagery. *Proceedings of IGARSS '87 Symposium*, Ann Arbor, Michigan, 18–21 May 1987, 353–356.

Laycock J (1990) Resampling algorithms for geocoding. *Proceedings of the Third International Workshop on Image Rectification of Spaceborne Synthetic Aperture Radar*, Farnham Castle, UK, 1–3 October 1990, 15–17.

Lee J-S (1981) Refined filtering of image noise using local statistics. *Computer Graphics and Image Processing*, **15**, 380–389.

Marr D and Hildreth E (1980) Theory of edge detection. *Proceedings of the Royal Society of London*, B, **207**, 187–217.

Raggam J (1990) SAR parameter adjustment and related modules for simulated ERS-1 data. *Proceedings of the Third International Workshop on Image Rectification of Spaceborne Synthetic Aperture Radar*, Farnham Castle, UK, 1–3 October 1990, 9–14.

Rignot E J M, Kwok R, Curlander J C and Pang S S (1991) Automated multisensor registration: requirements and techniques. *Photogrammetric Engineering and Remote Sensing*, **57** (8) August 1991, 1029–1038.

Schreier G (1991) SAR geocoding—operational service for ERS-1 and future sensors. *Proceedings of Spatial Data 2000*, Christ Church, Oxford University, 17–20 September 1991, 29–38.

Schreier G, Kosmann D and Roth A (1990) Design aspects and implementation of a system for geocoding satellite SAR-images. *ISPRS Journal of Photogrammetry and Remote Sensing*, **45**, 1–16.

Tilley D G and Bonwit K S (1989) Reduction of layover distortion in SAR imagery *Remote Sensing of Environment*, **27**, 211–220.

Weiler J, Brule M, Lim K, Ball R and Turner T (1986) An algorithm for automatically acquiring ground control points in SAR imagery. *Proceedings of the SAR Applications Workshop*, Frascati, Italy, 16–18 September 1986, ESA SP-264, 103–110.

3

Geocoded image products

This chapter describes how geocoded satellite images may be further enhanced and combined with existing map data to form a range of geocoded products. It contains the following sections:

- —product planning: definition of the required product in terms of location, format, and annotation;
- —image enhancement: the removal of instrument errors, contrast stretch, filter operations, and multiband operations designed to improve the visual appearance of the product;
- —production of satellite image mosaics: techniques for joining together of several geocoded images, improving the geometric and radiometric accuracy of the product and how the techniques were applied to the AVHRR mosaic of Antarctica;
- —image map annotation: adding digital map data, three-dimensional visualisation and the final production of the output product as hard copy.

A typical flowline for geocoded product generation can be seen in figure 3.1.

3.1 PRODUCT PLANNING

The first stage in preparing a geocoded product of a particular area is to assess the potential coverage by a particular satellite sensor. For example using the Landsat world reference system (WRS) a chart can be drawn up showing all the path/row combinations needed to cover an area of interest. Quick-look prints can then be used to select the best actual images to be used. The area of interest may conform to a country or correspond to a particular map sheet. The images should be as cloud-free as possible, although it is often possible to use one partially cloud-free image to patch in the cloudy areas on another. Ideally the images will be of similar appearance. If for example one image is snow-covered or images correspond to different seasons of vegetation growth it may prove difficult to join images in a seamless manner.

For many image maps (USGS 1986) it is preferable that growing vegetation is avoided, as the strong infra-red response may overwhelm cultural features. Often autumn imagery is the best—when deciduous leaf-cover has passed its growing season, or alternatively

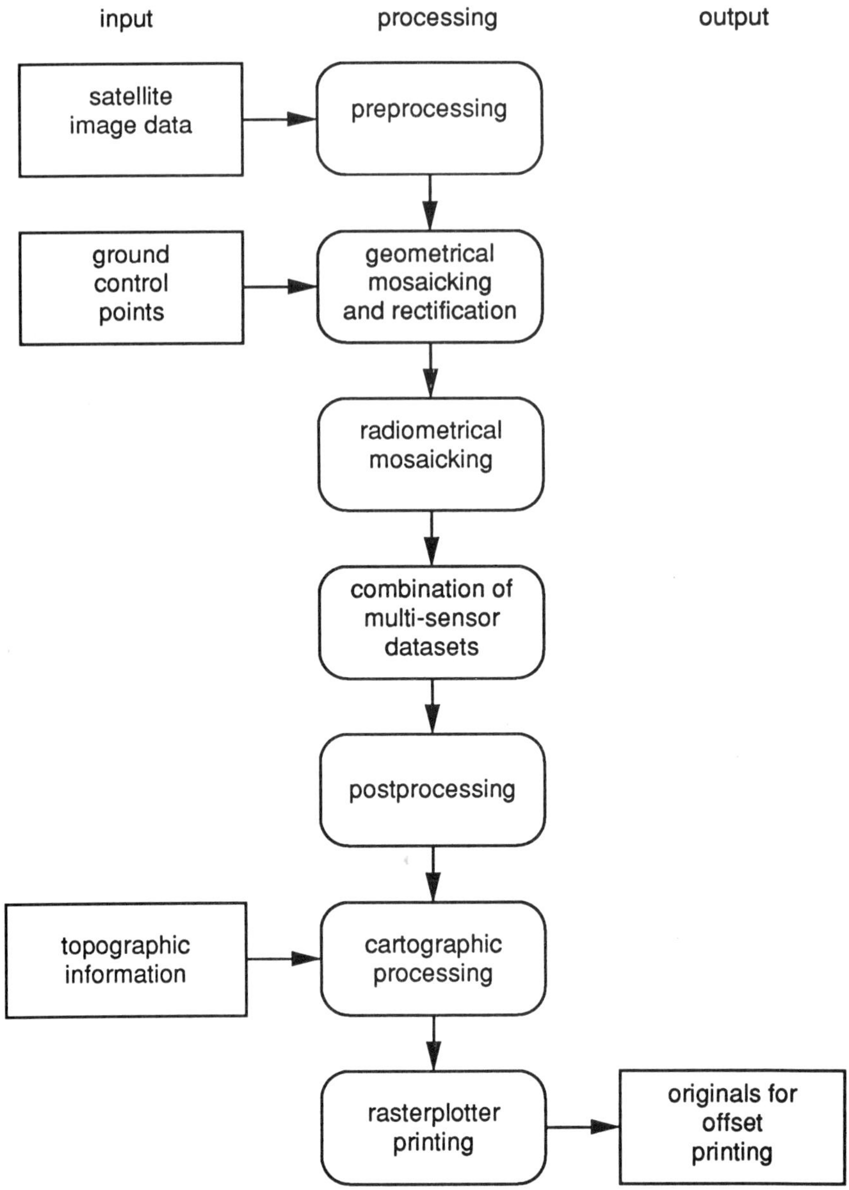

Fig. 3.1. Flowline for geocoded product generation (based on Albertz *et al.* 1992).

spring before the growing season has begun in earnest. The sun angle should not be too great, as a value of more than 50° will make it difficult to discriminate low-relief features due to lack of terrain shadows; it is also desirable that sun angles are roughly the same for each image to minimise the amount of radiometric correction necessary. A suitable sun angle will also enhance topographic structure, depending on the local relief. Table 3.1 shows appropriate values suggested by the US Geological Survey.

Table 3.1. Suitable sun angles for image maps (based on USGS 1986)

Terrain	Sun angle
Flat	0°–15°
Moderate	15°–20°
Steep	30°–50°

It is also necessary at this stage to decide on the grid size of the output image; because of the large amount of data involved, practical constraints on computer storage may dictate that this be larger than the pixel size of the input images. For example, a 1000 km square mosaic with a grid size of 25 metres (easily achievable with Landsat-TM) would constitute an image of 40 000 lines of 40 000 pixels requiring storage of 1600 megabytes of data. The storage required for a product will vary inversely with the square of the grid size. In the above example a grid size four times coarser (100 metres) would require one-sixteenth the storage space (100 megabytes) which may be a more manageable proposition.

At this stage it is usually advantageous to attempt one or more prototypes, for example using sub-sampled images which have been bulk geocoded to produce a 'mock-up' of the area being covered. This gives an indication of the final product which can be used to determine the location of 'cut lines' for the mosaic process. Based on this information a definitive project plan can be drawn up and used as a baseline for the product development (see figure 3.2).

3.2 IMAGE ENHANCEMENT

There are a variety of methods that can be used to enhance the visual interpretability of the image. Amongst the techniques commonly used in image map production are:

(1) removal of instrument errors—replacement of missing pixels or lines caused by defects in data acquisition and removal of the effects of sensor imbalance;
(2) contrast stretch—to improve the dynamic range of an image to make it easier to interpret visually;
(3) filter operations—the use of digital filters to emphasise the linear features often associated with important cultural objects (such as roads and rivers) or to reduce the effects of noise in an image;
(4) multiband operations—to choose a suitable colour combination of image bands, or to perform various enhancing transformations on multispectral data, including the merging of data from different image sources to maximise the particular information content of each.

HARRISBURG PROJECT PLAN

Objective

The objective of this project is to produce image map data for the Harrisburg 1:250 000-scale topographic area.

Production Overview

The Harrisburg quadrangle will be a 4-scene mosaic from Multispectral Scanner (MSS) data. Digital defects within the data will be identified and corrected using fixline software. Control point selection, cutline determination and mosaicking will be completed by Data Management personnel. Upon completion of the digital mosaicking procedure, image enhancement and final formatting will be completed by the Production Analysts with input from Data Management.

Input data

Bands utilized will be 4, 5 and 7

Scene IDs used in the project

83152115094X0	Path 16	Row 31	4 May 1982	Harr1631
83152115101X0	Path 16	Row 32	4 May 1982	Harr1632
83152215153X0	Path 17	Row 31	5 May 1982	Harr1731
83152215155X0	Path 17	Row 32	5 May 1982	Harr1732

Production parameters

- 50m pixels using cubic convolution resampling
- 60 micron spot size on Color Fire 240 recorder
- first generation scale = 1:833 333
- UTM projection (zone 18)
- insert quad corner tick marks into image data
- insert grey scales near top of image
- insert registration tick marks outside image area

Fig. 3.2. Sample project plan for a Landsat image map of Harrisburg, USA (from USGS 1986).

3.2.1 Removal of instrument errors

Data dropout

A common problem with satellite images is data dropout which may occur during image transmission from the satellite to a ground station or at a later processing stage (for example transcription onto magnetic tape). This may manifest itself as missing individual pixels, groups of pixels or even entire lines. The first stage in repairing these defects is to determine where the data is missing. This can be performed interactively using an IIPF or automatically using image statistics (for example searching for groups of zero values in otherwise bright areas). Line repair can then be performed either by substituting neighbouring pixel values or by interpolation from neighbouring pixels, for example replacing the missing pixel with the average of the pixel values above and below it, or with the average value of the three pixels above and three pixels below it. An alternative method of line repair is band substitution, where values are substituted from a similar band (for example Landsat-TM2 for TM1) either using unmodified values, or applying a contrast stretch based on the global statistics of the two bands. Alternatively values may be interpolated between two or more bands.

Striping

Another common instrument error is striping which occurs as follows. Most satellite sensor systems use a bank of adjacent sensors for each spectral band, for example six (Landsat-MSS) and sixteen (Landsat-TM). For push-broom sensors a much larger number of detectors is used to record an entire line at a time. Miscalibration between the detectors can lead to a striped effect which may appear as horizontal stripes (as in the case of Landsat), vertical (SPOT) or occasionally a combination of both. The simplest solution is to calculate the image histogram corresponding to pixels acquired with each sensor and to adjust the histograms by applying a simple linear contrast stretch (see section 3.2.2); this normalises the sensor output to that of a nominal median sensor. It is essential to perform this operation prior to the actual geocoding, which will (amongst other effects) rotate the stripes from the original grid pattern.

3.2.2 Contrast stretch

For many images only a small sub-range of the available dynamic range (0 to 255 for eight-bit data) are actually present in a particular image. When viewed an image will often appear dark and lacking in contrast and the human eye will not be able to discriminate the detail present in the image. The technique of modifying the image's dynamic range to suit the response of the human eye is known as contrast stretching. The simplest form of linear contrast stretch is to map the actual dynamic range onto the maximum available dynamic range. Thus if the actual dynamic range is $[I_{min}, I_{max}]$ this contrast stretch would be a linear mapping:

$$O(x) = \frac{255 \cdot [I(x) - I_{min}]}{I_{max} - I_{min}} \tag{3.1}$$

where

$$O(x) = \text{output image value,}$$
$$I(x) = \text{input image value,}$$
$$I_{\text{min}} = \text{maximum input value,}$$
$$I_{\text{max}} = \text{minimum input value.}$$

A representation of this contrast stretch in graphical terms as a 'transfer function' can be seen in figure 3.3.

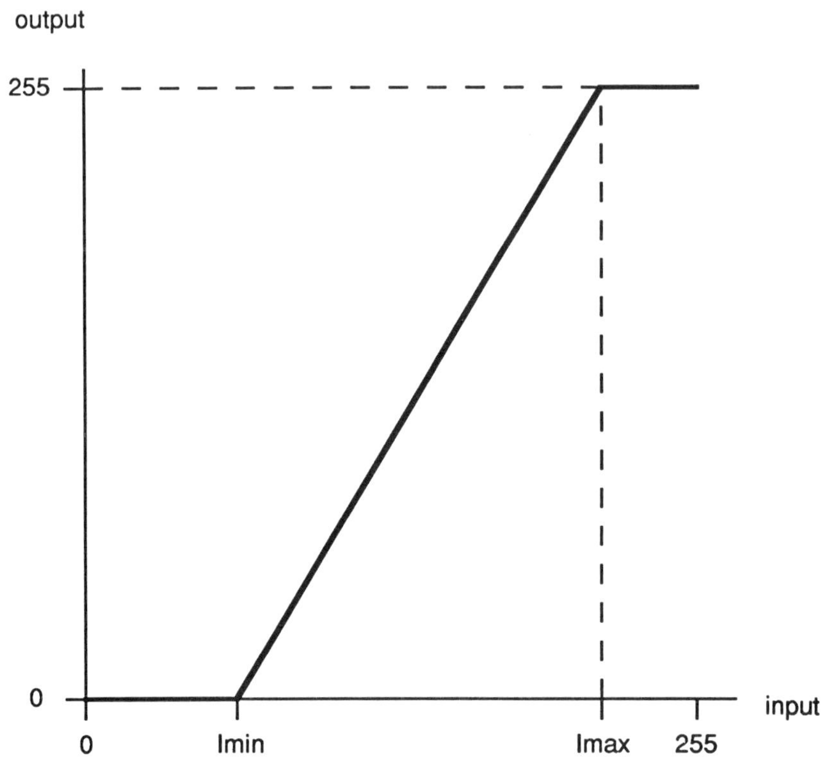

Fig. 3.3. Linear contrast stretch represented as a transfer function.

A contrast stretch can be usually performed using an IIPF by means of a look-up table (LUT). The LUT stores an array of numbers which determine the values displayed for each input value—by using a LUT the contrast stretch can be performed in real time without having to modify the actual input values. An example of the use of a LUT to store the transfer function can be seen in table 3.2. This shows a simple linear contrast stretch from an input range of [80, 131] onto the full dynamic range [0, 255], resulting in the transfer function shown in figure 3.4.

The simple linear contrast stretch may not prove satisfactory in the case where an image with an overall large dynamic range may have most of its information lying within

Table 3.2. Use of LUT to store a transfer function

Input value		Output value
0		0
1		0
2		0
	...	
79		0
80		0
81		5
82		10
83		15
84		20
85		25
	...	
129		245
130		250
131		255
132		255
	...	
254		255
255		255

a relatively small sub-range. This is typified by a fairly dark image with a few bright 'saturated' pixels, caused for example by highly reflective clouds or snow. One way to deal with this problem is to ignore the outlying values near the ends of the histogram, for example to only consider the data lying between the fifth and ninety-fifth percentiles. Another method is to use an IIPF to determine the end points for the stretch interactively.

A more complicated piecewise linear stretch can also be defined by specifying a series of (input, output) pairs. Any input value intermediate to two such points has an output value determined by linear interpolation (see figure 3.5). This type of contrast stretch may be used to give different emphasis to portions of the histogram corresponding to distinct terrestrial features for example sea, low-lying areas and snow-covered upland.

Other automatic contrast stretches are based on transforming the empirical distribution of an image histogram onto a standard histogram whose statistics are better understood. Distributions favoured include the gaussian (normal) and uniform distribution. In general better results will be achieved by such a non-linear contrast stretch. This is because the visual response of the human eye is non-linear, for example it requires a greater difference in density value near the extremes of black and white for a change to be perceivable, compared to the mid-grey regions.

3.2.3 Filter operations

Various image enhancements can be performed by means of operations with a filter kernel. A kernel is a matrix (often 3×3) containing a set of coefficients (or weights). The

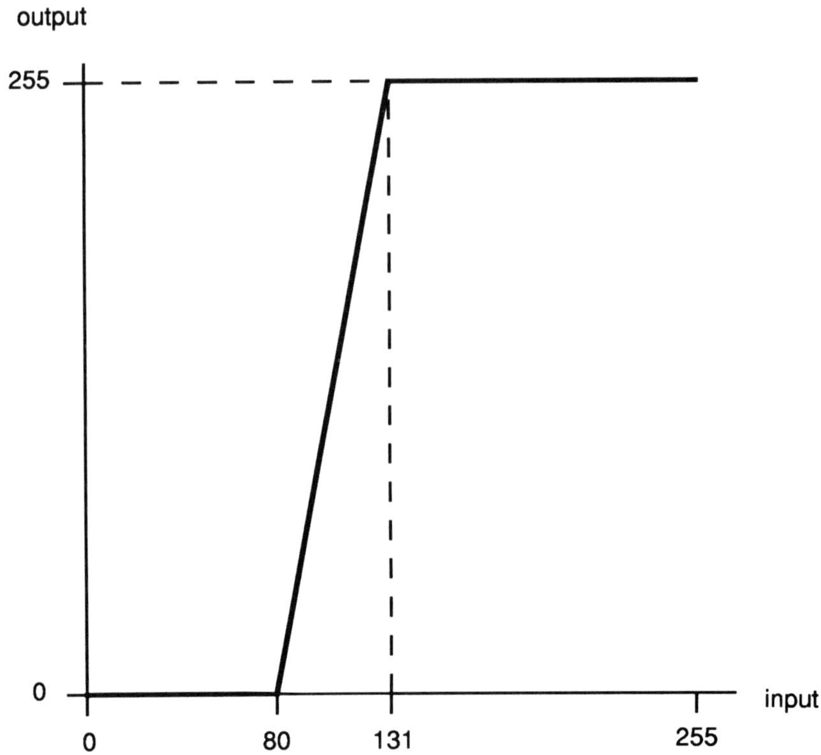

Fig. 3.4. The transfer function for the linear contrast stretch shown in table 3.2: the range [80, 131] is mapped onto the full dynamic range [0, 255]

value of each pixel in the output image is determined by placing the kernel over the corresponding pixel in the input image, multiplying all the pixels beneath the kernel by the appropriate weight and summing the result. The process, known as convolution, can be represented mathematically (for a 3×3 kernel) as follows:

$$O(x, y) = s \sum_{i=-1}^{i=+1} \sum_{j=-1}^{j=+1} K(i, j) \times I(x+i, y+j) + c \tag{3.2}$$

where

$O(x, y) =$ output pixel value,

$I(x, y) =$ input pixel value,

$s =$ scale factor,

$K(i, j) =$ filter kernel,

$c =$ constant offset.

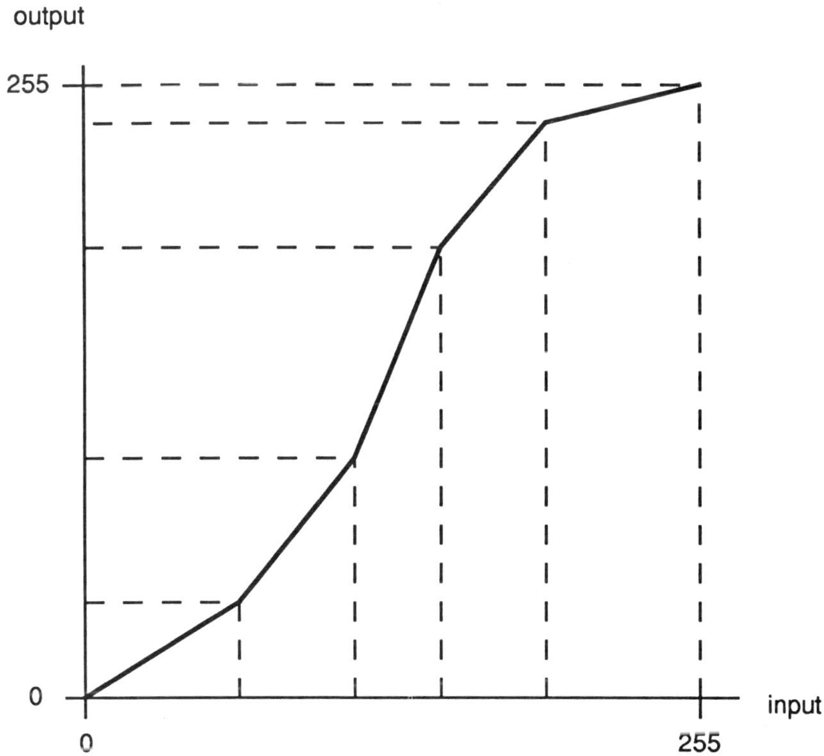

Fig. 3.5. Transfer function for a piecewise linear contrast stretch.

Note that the scale factor and offset are normally selected for convenience of display (for example to ensure an output in the range 0 to 255) and suitable values for a particular kernel can readily be found by experimentation using an IIPF. Convolution filters are normally classified as smoothing (or low-pass) filters and edge enhancing (or high-pass) filters.

Smoothing filters

Smoothing filters are used to remove noise in an image, which can produce unwanted high spatial frequency artifacts. The simplest low-pass smoothing filter is the mean filter whereby each pixel is replaced by the average value of its neighbours. In this case

$$K = \begin{bmatrix} 1 & 1 & 1 \\ 1 & 1 & 1 \\ 1 & 1 & 1 \end{bmatrix} \qquad s = \frac{1}{9}.$$

For isolated 'salt-and-pepper' noise a non-linear median filter, whereby each pixel is replaced by the median value of its neighbours may be preferable to a linear filter as it ignores extreme high or low values, rather than blending them in. (The median filter is strictly speaking, however, a neighbourhood operation rather than a convolution filter.)

Edge enhancing filters

Most of the information of interest in an image is derived by the human eye from the edges, that is, the areas where there are sudden changes in the intensity of the image. Several algorithms can be used to enhance these edges to make them more apparent visually. The skill in applying a suitable algorithm is to select one which will emphasise the genuine edges of interest without creating artifacts from spurious edges, for example, caused by noise or data dropout. A general approach to enhancing edges in an image is to emphasise the difference between the central pixel and its neighbours. For example the laplacian kernel

$$K = \begin{bmatrix} -1 & -1 & -1 \\ -1 & 8 & -1 \\ -1 & -1 & -1 \end{bmatrix}$$

produces a zero result in homogeneous areas (those with little variation in pixel values) and a large positive or negative result when the pixel is significantly higher (or lower) in value than its surroundings.

By blending a suitable amount of the laplacian output in with the original image an enhanced version is achieved.

$$O(x,y) = a \times I(x,y) + b \times L(x,y) \tag{3.3}$$

where

$$\begin{aligned} O(x,y) &= \text{ output pixel } (x,y), \\ I(x,y) &= \text{ input pixel } (x,y), \\ L(x,y) &= \text{ laplacian convolution of } I(x,y), \\ a,b &= \text{ constants.} \end{aligned}$$

Edge enhancing filters can also be directional, for example the Sobel filters, K_1 and K_2

$$K_1 = \begin{bmatrix} -1 & -2 & -1 \\ 0 & 0 & 0 \\ 1 & 2 & 1 \end{bmatrix} \qquad K_2 = \begin{bmatrix} -1 & 0 & 1 \\ -2 & 0 & 2 \\ -1 & 0 & 1 \end{bmatrix}$$

emphasise horizontal and vertical edges respectively.

It is also possible to combine edge enhancement filters with smoothing filters. For example (Kidwell and McSweeney 1984) the following technique can be used to magnify an image:

— magnify the image by a factor of two using pixel duplication;
— smooth the image using a low-pass filter;
— perform edge enhancement using a high-pass filter.

This method avoids the noticeable blocked appearance found when pixel replication is used on its own. The best results are usually found (in the case of Landsat-TM data) by using an edge enhancing kernel twice as large as the multiplication factor (e.g. 4×4 filter for a factor of 2 multiplication).

3.2.4 Multiband operations

Band selection

There are two basic colour definitions for image maps; either monochrome or some kind of colour combination. If a multispectral sensor such as Landsat-TM or SPOT is used as the basis for a monochrome image map it is usual to select an infra-red band (such as TM7) as this usually provides better feature discrimination. Alternatively, a panchromatic sensor such as SPOT-PA may be used. A colour image map on the other hand will use the sensor's ability to image an area in several portions of the spectrum, not only visible but into the infra-red. On a display device, or a photograph, three image bands (one for each of red, green and blue) can be depicted. A SPOT-XS image with exactly three bands can be mapped exactly onto the three available display bands. With images possessing more than three bands, several methods can be used to reduce the number to three. With some images it is possible to produce a natural colour image. For example with Landsat-TM there are three visible wavebands located approximately at the three primary colours. Thus the combination:

$$\{3, 2, 1\} \to \{R, G, B\}$$

will produce an image with similar properties to a standard colour photograph of the scene which is simple to interpret as it corresponds to familiar colours for example green fields. There are of course many other possible combination—ignoring the thermal band (TM6) because of its larger pixel size, there are 20 combinations of three bands from six and 120 permutations when they are mapped onto the primary colours. From the multitude of possibilities a few have proved most suitable (Dowman and Peacegood 1988) for the production of image maps, for example:

$$\{5, 3, 2\} \to \{R, G, B\}.$$

This has proved useful where good water body discrimination is required. However, the most suitable combination for an image map will usually depend on the predominant land cover in the image. For example, in areas of dense vegetation band 4 has a characteristically high response so summer scenes, with large areas of vegetation, may appear unbalanced with towns appearing swamped by surrounding countryside; in this case band 5 which has a lower vegetation response, may prove a more suitable substitution.

It is also possible to select the best selection of bands mathematically (Chavez *et al.* 1984; Sheffield 1985). This is achieved by examining the variance–covariance matrix for the band combinations of a specific image. The variance of a band gives a measure of the amount of information within a band, whereas the covariance between two bands measures how closely the information contained in the first band resembles that in the second. Ideally therefore the three bands selected should have large individual variance, but there should be little correlation between any two of them. The best band combination can be found by taking the variance–covariance matrix $M(i, j)$ and calculating the determinants of each principal sub-matrix—the original values for the thermal band are scaled by a factor of 4 to account for the coarser resolution of 120 m. There are 35 principal (3×3) sub-matrices corresponding to each possible three-band combination. The triplet $B(i, j, k)$ with the maximum value is the best one to choose but there are still six permutations that map it onto {R,G,B}. The selection is made by examining the variance in each band (the diagonal entries in M). The eye is most sensitive to green, then to red, then to blue; so the band with most variance is assigned to green, second to red and third to blue. The selected band combination will again depend on the area being imaged but the algorithm has been found to generally pick out combinations with only one visible band whereas human operators tend to prefer combinations with two.

IHS transform

Another approach to making image maps is to combine imagery from two sources in a way that utilises the best attributes of both of them, for example the high-resolution of a SPOT panchromatic image with the spectral coverage of Landsat-TM. One useful tool in the merging of optical imagery is the intensity hue saturation (IHS) algorithm. This is particularly suited for the combination of multispectral imagery such as SPOT-XS or Landsat-TM with higher-resolution monochrome images such as SPOT-PA, to provide a simulated high-resolution multispectral product (Carper *et al.* 1990). A simplified portrayal of IHS space can be seen in figure 3.6.

The IHS transform allows manipulation of spatial information without destroying overall colour balance. The intensity is related to the overall brightness of a pixel, the hue to its colour (that is the dominant spectral wavelength) and saturation to its purity

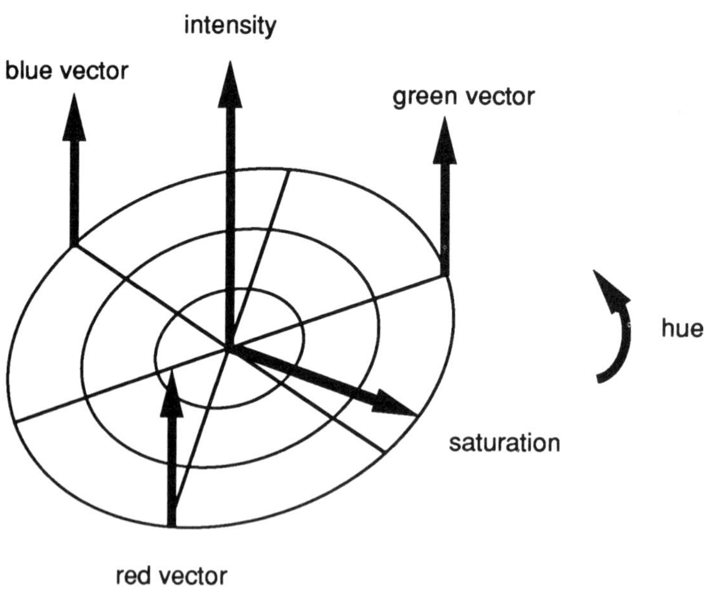

Fig. 3.6. Representation of colours in an IHS coordinate system (adapted from Carper *et al.* 1990)

(relative to grey). A suitable procedure for merging a high-resolution panchromatic image and a lower-resolution multispectral image is shown in figure 3.7. The first step is to geocode both images and then to transform the low-resolution image from (R,G,B) to (I,H,S) space. It is also necessary to resample the low-resolution image to the

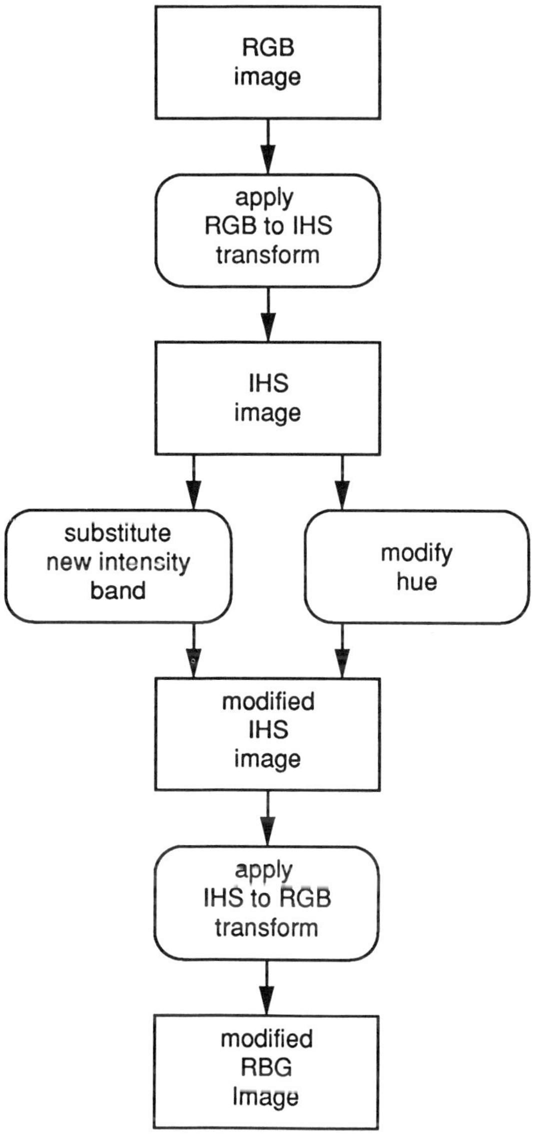

Fig 3.7 Merging high-resolution panchromatic images with lower resolution multispectral images using the IHS transform—the panchromatic image is substituted for the intensity band, the 'modify hue' stage is optional, for example to improve the colour balance.

higher-resolution, using for example a cubic convolution technique (Welch and Ehlers 1987) or by pixel replication. At this stage the panchromatic image is substituted for the intensity component prior to transformation back from (I,H,S) to (R,G,B). The result is an image that has the high resolution of the panchromatic imagery but retains the spectral characteristics of the multispectral imagery.

It is also possible to perform additional processing in the (I,H,S) space before conversion back to (R,G,B) space, for example:

— histogram equalisation of the hue component to improve the colour balance;
— rotation in colour space to move histogram peaks to the desired colour (for example, to produce a greener image).

3.3 PRODUCTION OF SATELLITE IMAGE MOSAICS

3.3.1 Mosaicking techniques

Where the geographic cover of an image map spans several component images, they can be joined to produce a satellite mosaic. The production of such a mosaic (Zobrist *et al.* 1983) is a skilled and labour-intensive process, with much intervention from a trained operator required, but the result is often a visually attractive product as well as a valuable source of cartographic information unobtainable by conventional means. Examples of satellite image mosaics are radarmaps of Germany and Antarctica being prepared using ERS-1 SAR data (Kosmann *et al.* 1992) and the mosaics of Europe (see figure 3.8 in the plate section between pp. 84 and 85) and Antarctica (see section 3.3.3 and the front cover) produced by the UK National Remote Sensing Centre.

The production of an image mosaic can require an immense amount of data—an idea of the effect of data type and hence grid size on the output product size can be seen in table 3.3, showing the coverage required for a 1000 km square area.

Table 3.3. Output product size as a function of input data type

Sensor	Grid size (m)	Mosaic size	Storage (megabytes)
AVHRR	1000	1000 × 1000	1
Landsat-MSS	100	10 000 × 10 000	100
Landsat-TM	25	40 000 × 40 000	1600
SPOT-PA	10	100 000 × 100 000	10 000

A common approach to mosaic production is to join one image at a time onto the partially completed mosaic. Thus the first step is to join two overlapping images together (see figure 3.9).

In order to balance the radiometric appearance of the component images a linear contrast stretch is applied so that the mean and variance of the overlap region in both images is the same. It can be seen that any point on the mosaic grid falls into one of four categories:

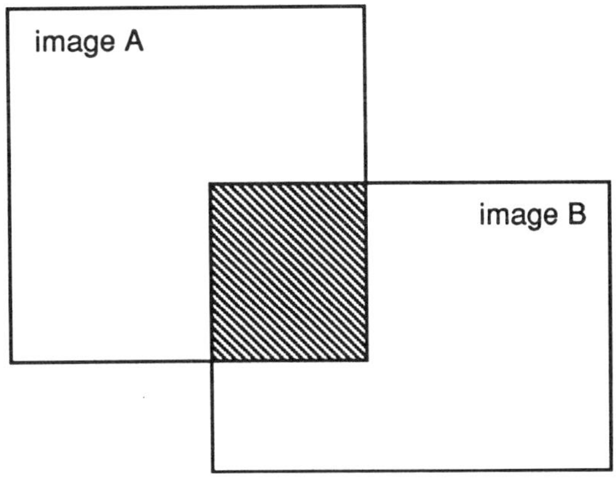

Fig. 3.9. Joining two overlapping images A and B (the overlap area is shaded).

(a) in the first image only;
(b) in the second image only;
(c) in neither image;
(d) in both images.

In cases (a) and (b) the output pixel in the mosaic is set to the value of the corresponding input pixel. In case (c) it is set to a background value (it is customary in mosaicking practice to reserve a value such as 255 as a background value—genuine data with a value of 255 can be reset to 254 with little loss of significant radiometric information). In the overlap area, various simple algorithms can be used to calculate the overlap value; the most often used ones are 'brute force' (set to the value of the first image, ignoring the second), calculate the average value of the two pixels or use the maximum (alternately minimum) of the two values.

The disadvantages of the simple approach is that the algorithms used can tend to create an artificial seam at the edge of the overlap region. It is desirable therefore to smooth out the values in the intersection so that the output value is $A(i, j)$ if (i, j) is on the seam of A, $B(i, j)$ on the seam with B, and a smoothly weighted blending function at intermediate points. A suitable method is based on the mathematical measure of distance known as the taxi-cab metric. This is a computationally efficient method of defining the distance between two points as the sum of the difference in pixel numbers and the distance in line numbers as follows:

$$d[(x_1, y_1), (x_2, y_2)] = |x_1 - x_2| + |y_1 - y_2| \tag{3.4}$$

where

$(x_1, y_1) =$ position of first point,

$(x_2, y_2) =$ position of second point.

This can be used as the basis of a function z which measures the relative closeness of a point from A and B on a scale of [0, 1] such that $z = 0$ on the join with A and $z = 1$ on the join with B and varies smoothly between them. In terms of the taxi-cab metric:

$$z = \frac{d_a}{d_a + d_b} \tag{3.5}$$

where

$z =$ relative closeness function,

$d_a =$ distance from nearest point on join with A,

$d_b =$ distance from nearest point on join with B.

The function z can then be used to define the weighting function $w(z)$ used to define the output value:

$$M(i,j) = w \times A(i,j) + (1 - w) \times B(i,j) \tag{3.6}$$

where

$M(i,j) =$ mosaic value at point (i,j),

$A(i,j) =$ value at point (i,j) in image A,

$B(i,j) =$ value at point (i,j) in image B.

Once the first two images have been joined, the same procedure can be repeated for each subsequent image by regarding the partially completed mosaic as 'Image A'. The merging of images is performed by assembling the component geocoded images on a regular grid. Best results are obtained by merging images acquired on the same along-track pass followed by adjacent passes. At this stage any filler data is replaced by genuine image data. A less obvious join between two images can be achieved if, instead of using straight line boundaries, natural features such as field boundaries are used. The first stage is to specify the join line between the two images; there are two main ways of doing this:

(1) using an IIPF to draw in the join line;
(2) using the IIPF to specify a series of tie points along the desired boundary and use a suitable interpolation method (e.g. linear, cubic spline) to join and complete the line.

It is then a simple matter to process each point in the overlap and to allocate a value depending on which side of the boundary line it falls. This was the method employed by the UK National Remote Sensing Centre to produce a series of regional mosaics from Landsat-TM data (Batts, personal communication). Where a mosaic encompasses a coastal region it is necessary to mask out the sea to remove the great variation in the sea's appearance from image to image which cannot be removed by blending algorithms. The sea mask may be obtained by using an IIPF or by automatically defining the sea boundary by density slicing or histogram thresholding. Care must be taken also with any remaining areas of cloud—it is undesirable to blend in cloud values with genuine surface features. It is difficult, however, to reliably remove clouds automatically, so a skilled operator will paint out clouds using an IIPF.

3.3.2 Improving mosaic accuracy

Improvements in mosaic accuracy fall into two classes—radiometric and geometric. One commonly used method of radiometric correction is to equate the histograms of the two images within the overlap area. This is normally achieved by using the cumulative histograms of the two images and adjusting them so that they agree at selected percentile points. This can be achieved by use of a LUT, interpolating between points as necessary. Another method (Binnie and Colvocoresses 1987) is to perform radiometric matching using a sampling method. Image patches (size 9×9) are analysed along the seam between two mosaic components to compare the average brightness value between the two images; this information is then used as input to a piecewise linear surface fit between the two sets of radiometric values.

A set of registration tie points in the overlap area can be used to produce a more accurate geometric correspondence where the two images join. This method is used, for example, to automate the production of mosaics from SAR images (Kwok *et al.* 1990). Frame-to-frame registration in the overlap region is performed by selecting image patches and performing one of the following pattern matching operations:

— cross-correlation method;
— zero crossing method.

The zero crossing method is based on the Marr–Hildreth algorithm (Marr and Hildreth 1980) and may be preferable because changes in SAR image geometry and radiometric characteristics (such as the multiplicative noise evident as speckle) tend to decorrelate the image. A similar method was used for the production of mosaics from Landsat-TM images (Binnie and Colvocoresses 1987) with the refinement that tie points in the seam region are triangulated, and a transformation is defined on each of the triangles in order to form a continuous model for the entire image map.

3.3.3 The AVHRR mosaic of Antarctica

One of the most impressive satellite mosaics is the AVHRR mosaic of Antarctica produced by the UK National Remote Sensing Centre (Merson 1989). The raw material for the mosaic consists of a set of 50 NOAA-AVHRR images of the continent which vary greatly in data integrity and cloud cover. The best 25 scenes were selected to form the relatively cloud-free basis of the mosaic. AVHRR bands 1,2 and 4 were selected to give a consistent colour scheme throughout. Before the individual images were used they were compressed from their original 10-bit dynamic range [0, 1023] down to 8-bits [0, 255] for subsequent processing, and any data dropouts were replaced by a low-pass line repair algorithm. The images were acquired illuminated by a variety of sun angles. This led to considerable radiometric inequalities within the image set as the reflectance in bands 1 and 2 tend to be highly correlated with the sun angle. Because of the large area covered by an AVHRR image it is not reasonable to assume that the sun angle remains almost constant across the width of an image (as is the case with Landsat), thus a project-specific algorithm was devised to correct for sun angle effects.

The first stage of processing for each image consisted of producing a sun angle image derived from sun–sensor–target geometry, with pixel values proportional to the sine of the sun elevation. An IIPF was then used to produce scatter diagrams of the sun image pixel against the intensities from the actual image. The best-fit quadratic curve to the scatter diagram was then used to produce a numeric correction to the raw intensity levels over the whole scene. A study of the map projections commonly used for world mapping reveals that many of them (for example Mercator) are not suitable for mapping polar regions, typically distorting regions of high latitudes or even disconnecting them completely. For this reason the projection used was polar stereographic (centred on the south pole). Each image was geocoded separately. A first approximation was made using satellite orbital ephemeris and scanner parameters to correct for earth curvature and rotation and variations in satellite orbit and attitude. Another problem with images of the Antarctic is the almost complete lack of suitable accurately located GCPs. This was circumvented by obtaining a digitised Antarctic coastline from the UK Scott Polar Research Institute (SPRI), consisting of 222 000 points compiled from various sources between 1961 and 1981. An extract was taken from the coastline and transformed to raster format for each image to be transformed. This allowed tie points to be identified on the coastline and on the actual image. The locations of the points in each coordinate system were then stored and the best-fit bilinear matrix calculated to model the transformation. This was then used to refine the approximate transformation derived from the ephemeris data. The final step was a cubic convolution resampling.

The method used for joining images was slightly different from the overlapping method described in section 3.3.1. In this case, the mosaic was constructed using abutting images, that is, every pixel in the final output mosaic was derived from a single input image. This method was preferred because the large radiometric variations in the input images made the algorithms normally used for blending overlapping images unsuitable. To record the final positions of the constituents of the mosaic a master mask was constructed. This is a grid of pixels corresponding to the size of the final mosaic. Each pixel value is a code indicating which input image will be used to produce the corresponding 1 km square (see figure 3.10).

The master mask was then used to blank out the pixels in each component which are not going to be used. The scenes used for the final mosaic are shown in table 3.4.

Various techniques were used to complete the final mosaic:

(1) adjacent components were balanced using a linear contrast stretch to minimize apparent seams;
(2) a smoothing algorithm was applied to get rid of any remaining discontinuities;
(3) areas of sea were masked out and set to a uniform value.

The sea masking was necessary due to the large variations in sea state and ice margin. The completed mosaic, which can be seen on the front cover of this book, provides a visually striking, and scientifically useful product with a variety of potential applications in geological and other polar research.

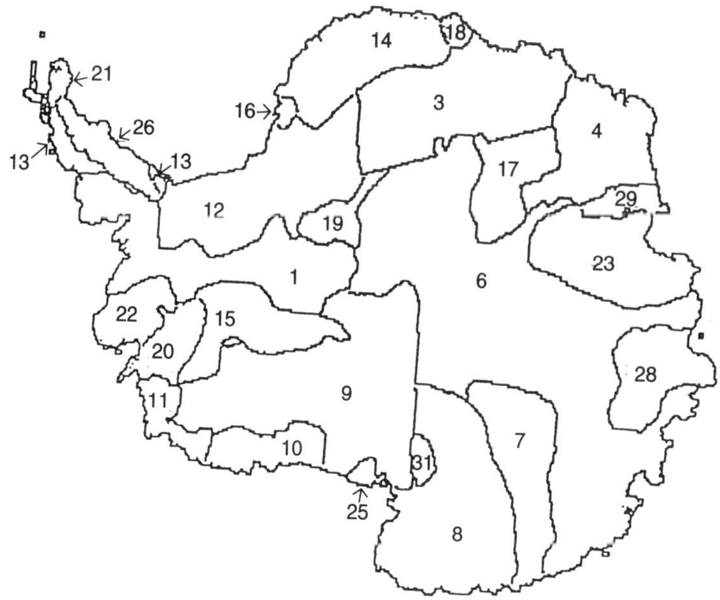

Fig. 3.10. The master mask used for the construction of the AVHRR mosaic of Antarctica (from Merson 1989).

3.4 IMAGE MAP ANNOTATION

3.4.1 Adding map data

The two main options for adding map data are to either treat the geocoded product as a standard photographic product and add map information as analogue line/lettering separates at the printing stage, or by superimposing it digitally. The use of wholly digital techniques permits a far wider range of products and allows experimental or prototype products to be generated with a rapid turnaround time. Features of interest that may be superimposed on a geocoded product include urban areas and communications such as road, rail and air routes. Annotation can include the names of conurbations and major natural features such as mountains, lakes and rivers; political and administrative boundaries; and map grids to provide scale and orientation information.

As with conventional map data, height information can be added in the form of spot heights or contours. Another use of map data is to use an image mask to highlight areas of interest on an image map, for example a mask could be used to define an administrative boundary. Within the boundary the image map is enhanced to bring out features of interest; outside the boundary it is subdued to provide background and continuity information. Another alternative is to incorporate the map data as a raster by scanning the entire map into an image format which (since it will already be in a standard map projection) can be easily combined with a geocoded image.

With digital map data in vector format (see section 6.1.2) a certain amount of format conversion is necessary. The basic technique is to consider the image as a grid onto

Table 3.4. AVHRR Antarctic scenes used to produce mosaic (based on Merson 1989)

Date	Satellite	Longitude (°E)	Mask key
12 January 1980	NOAA 6	148–184	8
13 January 1980	NOAA 6	230–280	22
14 January 1980	NOAA 7	74–140	6
18 January 1980	NOAA 6	62–121	23, 29
25 January 1980	NOAA 6	246–324	13, 20
11 February 1980	NOAA 6	298–335	12
15 February 1980	NOAA 6	42–67	4
6 November 1980	NOAA 6	210–270	11
8 November 1980	NOAA 6	290–304	15, 26
12 November 1980	NOAA 6	168–230	9
24 November 1980	NOAA 6	280–294	1
27 November 1980	NOAA 6	330–31	16–19
24 January 1981	NOAA 6	300–310	21
29 January 1983	NOAA 7	350–39	3
6 February 1983	NOAA 7	333–10	14
12 February 1983	NOAA 7	140–220	10, 25, 31
16 February 1983	NOAA 7	57–116	28
29 November 1983	NOAA 7	143–172	7

which the vectors making up the digital map data are drawn. The most important parameters of the grid are the four corners of the image in map coordinates and the pixel size (which defines the grid separation). The computer graphics operations applied include:

(1) scaling and translation—conversion of the map data to the same size and position as the image;
(2) windowing—the selection of vectors lying within the image grid;
(3) clipping—the calculation of the intersection of vectors with the grid boundary.

The actual assignation of overlay values is performed by determining the subset of grid squares that a particular vector passes through, these are then set in a map overlay image. In the simple case where a single map feature (e.g. roads, administrative boundaries) is to be overlaid then a binary image format is used with a value of 1 being used to indicate the presence of the feature within a grid cell and a value of 0 its absence (see figure 3.11).

In more complicated situations a feature coded representation is employed where the pixel value, instead of representing a physical value (image intensity) is a code representing the class of map feature, for example:

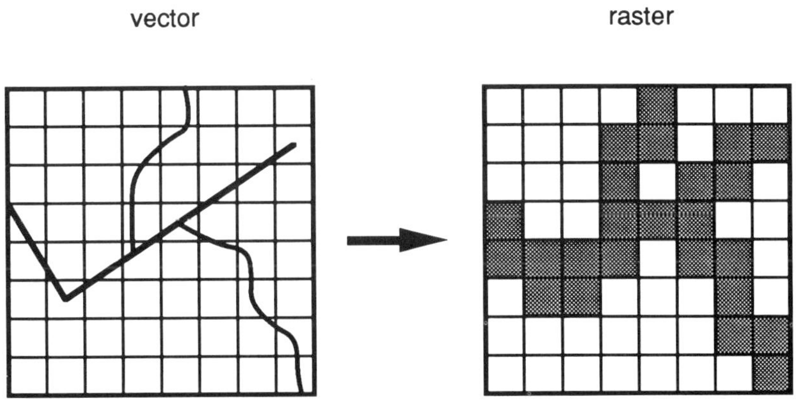

Fig. 3.11. Vector to raster conversion for a binary overlay.

0 = background

1 = minor road

2 = major road

3 = motorway

4 = railway

5 = canal

10 = other communication.

For display purposes the feature values are converted to a unique colour on the graphics overlay. An example of the use of SPOT panchromatic data, together with overlaid vector data to produce an image map of San Diego, California, can be seen in figure 3.12 in the plate section between pp. 84 and 85.

3.4.2 Three-dimensional visualisation

One of the most useful ways of displaying a geocoded product is to use an IIPF to drape it over a DEM. Figure 3.13 in the plate section between pp. 84 and 85 shows a perspective view of Sodaville, Nevada, USA. It comprises a Landsat-TM image superimposed over a DEM of the area, which was also created from Landsat-TM images using the techniques described in section 4.2. IIPFs are now becoming powerful enough to allow the viewpoint to be moved in real time allowing the effect of a moving observer or a fly-through to be constructed.

The use of satellite images as a backdrop to information in a geographic information system (GIS) can prove a very powerful method for viewing the appearance of terrain. For example a study carried out to assess the environmental impact of a proposed new town development in the Hart District of the UK (Quarmby *et al.* 1988) used a conventional vector-based GIS. No satellite images were used in the GIS itself which made use of digital maps of land use and proximity maps showing potential accessibility to motorways, stations and major roads. The land cover maps were used to derive information regarding the constraints on development for example existing built-up areas, airfields and protected land. The GIS also incorporated parameters which could be adjusted to model, for example, the relative importance of existing common land or areas of ecological or agricultural significance. The output from the GIS was a selection of potential development areas of varying suitability. It is at this stage that satellite images are used—SPOT images of a potential area are draped over a DEM which can then be used to provide perspective views of the development site from any desired viewpoint. The display can then be further augmented with computer graphics of proposed buildings, to produce a more realistic simulation of the situation of a proposed new town than could be provided either by the GIS output, or by traditional photo-montage techniques.

3.4.3 Hard copy

The final output product is most useful if produced at a standard mapping scale. The general rule of thumb (Colvocoresses 1986) is that the pixel size should equate to that of 0.3 mm in the final hard copy product. Thus with Landsat-TM a pixel size of 30 m would render it a suitable source for mapping at a scale of about 1:100 000 and SPOT-PA with a pixel size of 10 m somewhere between 1:25 000 and 1:50 000. A summary of suitable scales for satellite image mapping can be found in table 3.5.

Table 3.5. Appropriate scales for image maps (based on Albertz *et al.* 1992)

Sensor	Pixel size (m)	Scale
Landsat-MSS	80	1:250 000
Landsat-TM	30	1:100 000
SPOT-XS	20	1:100 000
SPOT-PA	10	1: 50 000

Although, as has been demonstrated previously in this chapter, there has been a considerable body of theory to support the production of geocoded products, ultimately a considerable amount of technical and artistic skill is necessary. As Kidwell and McSweeney of the US Geological Survey describe it:

> '... *image maps are both science and art in one medium*'
>
> (Kidwell and McSweeney 1984)

One important criterion for the production of hard copy geocoded products is the range of colours available (the colour gamut). In fact three colour gamuts are of paramount importance (see figure 3.14):

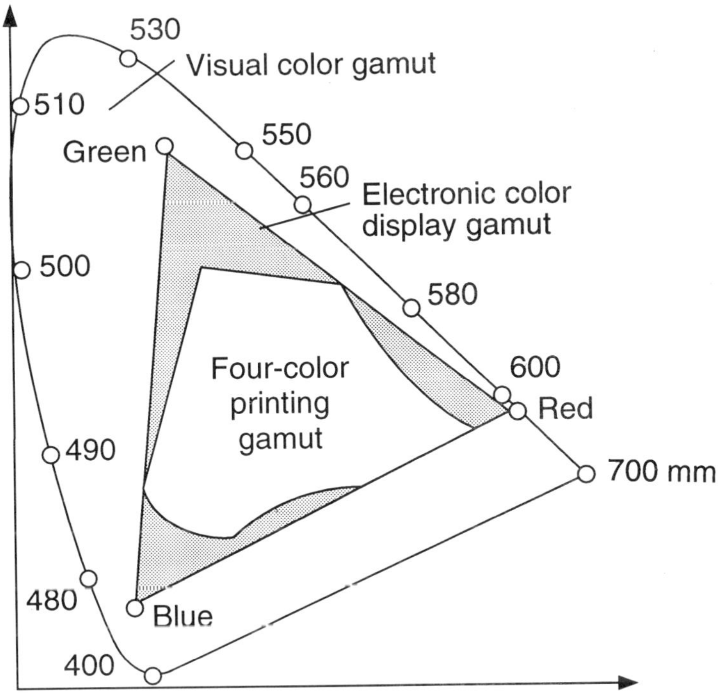

Fig. 3.14. Comparison of visual colour gamut with electronic colour display gamut and four-colour printing gamut (courtesy USGS; adapted from Williamson 1983).

— the visual colour gamut;
— the colour visual display gamut of an IIPF;
— the colour printing gamut.

As can be seen more colours are available in the 'real life' gamut than on the display of an IIPF, which in turn encompasses more colours than the colour printing gamut. This should be borne in mind when the beautifully balanced colour image on an IIPF produces a disappointing hard copy product! When printing a geocoded product the subtractive primary colours cyan, magenta and yellow are used instead of the primary additive colours (red, green and blue respectively). Guidelines for successful visual appearance (Kidwell and McSweeney 1984) include:

(1) look for good detail and contrast on the cyan separation;
(2) make sure that the magenta and yellow separations map well onto the cyan separate producing clear, bright colours;
(3) check that spectral highlights and white clouds are free of colour;
(4) check that dark shadow is black or dark grey;
(5) check for good contrast in the mid-tones.

Hard copy image maps are usually produced by using standard printing/photo-processing methods (for example generating a photographic negative from the digital data by film-writing) or by direct output on a hard copy device such as an electrostatic plotter.

3.5 REFERENCES

Albertz J, Lehmann H and Tauch R (1992) The production of satellite image maps—experiments at the Technical University of Berlin. *Proceedings of Satellite Symposia 1 and 2: Navigation and Mobile Communications, and Image Processing, GIS and Space-assisted Mapping, from the 'International Space Year' Conference*, Munich, Germany, 30 March—4 April 1992, ESA ISY-2, 175–180.

Binnie D R and Colvocoresses A P (1987) The Denali image map. *Photogrammetric Engineering and Remote Sensing*, **53** (3) 307–310.

Carper W J, Lillesand T M and Kiefer R W (1990) The use of intensity-hue-saturation transformations for merging SPOT panchromatic and multispectral image data. *Photogrammetric Engineering and Remote Sensing*, **56** (4) 459–467.

Chavez P, Guptill S C and Bowell J A (1984) Image processing techniques for Thematic Mapper data. *American Society of Photogrammetry, 50th Annual Meeting*, Washington DC, 11–16 March 1984, 728–743.

Colvocoresses A P (1986) Image mapping with the Thematic Mapper. *Photogrammetric Engineering and Remote Sensing*, **52**, 1499–1505.

Dowman I J and Peacegood G (1988) Information content of high resolution satellite imagery. ISPRS Congress, Kyoto.

Kidwell R D and McSweeney J A (1984) Art and science of image maps. *Proceedings ASPRS 51st Annual Meeting*, 10–15 March 1984, Washington DC, Vol 2, 771–782.

Kosmann D, Roth A, Schreier G and Winter R (1992) The ERS-1 radarmaps of Germany and the Antarctic peninsula. *Proceedings of the Central Symposium of the 'International Space Year' Conference*, Munich, Germany, 30 March–4 April 1992, ESA ISY-1, 321–325.

Kwok R, Curlander C and Pang S S (1990) An automated system for mosaicking spaceborne SAR imagery. *International Journal of Remote Sensing*, **11** (2) 209–223.

Marr D and Hildreth E (1980) Theory of edge detection. *Proceedings of the Royal Society of London*, B, **207**, 187–217.

Merson R H (1989) An AVHRR mosaic image of Antarctica. *International Journal of Remote Sensing*, **10** (4 and 5) 669–674.

Quarmby N A, Cushnie J L, Smith J (1988) The use of remote sensing in conjunction with Geographic Information Systems for local planning. *Proceedings of IGARSS '88 Symposium*, Edinburgh, Scotland, 13–16 September 1988.

Sheffield C (1985) Selecting band combinations from multispectral data. *Photogrammetric Engineering and Remote Sensing*, **51** (6) 681–687.

USGS (1986) *Procedure manual for preparation of satellite image maps*, National Mapping Program Technical Instructions, Open-File Report 86-19, Department of the Interior, US Geological Survey, National Mapping Division.

Welch R and Ehlers M (1987) Merging multiresolution SPOT HRV and Landsat TM data. *Photogrammetric Engineering and Remote Sensing*, **53** (3) 301–303.

Williamson S J (1983) *Light and colour in nature and art*, John Wiley and Sons, New York.

Zobrist A L, Bryant N A, and McLeod R G (1983) Technology for large digital mosaics of Landsat data. *Photogrammetric Engineering and Remote Sensing*, **49** (9) 1325–1335.

4

Global topographic mapping from space

One of the most important data structures used for spatial analysis is a digital elevation model (DEM) which is a digital representation of the earth's topography, for example a matrix containing the height of points on the earth's surface above some reference datum. Unfortunately, DEMs of usable detail are still not available for much of the earth's surface, and even when available may not be of sufficient accuracy. DEMs have great importance in the geocoding of both optical and SAR images, in the calibration of images (for example to provide an input to a reflectance or backscatter model) and in the visualisation of images as well as a wide variety of other applications. Faced with the increased global demand for DEMs satellite data provides a realistic source for all but the finest grid DEMs.

Chapter 4 describes topographic mapping from space and contains the following sections:

— photogrammetry using satellite images: the SPOT geometry that makes it suitable for derivation of height information, alternative along-track geometries and photogrammetric methods of height extraction;
— digital stereo matching: correlation, feature-based and hybrid stereo matching methods and the validation of the DEM produced;
— production of DEMs from SAR: the three main methods of processing stereo SAR (radargrammetry, interferometry and shape from shading);
— applications of DEMs: the importance of DEMs as an adjunct to information extraction from satellite images and for a variety of other application.

4.1 PHOTOGRAMMETRY USING SATELLITE IMAGES

4.1.1 SPOT stereo geometry

The most common method used to derive DEMs from satellite images is to use the parallax information present in a pair of images viewing the same scene at different angles. This is similar to the way that the human visual system obtains three-dimensional information (see figure 4.1).

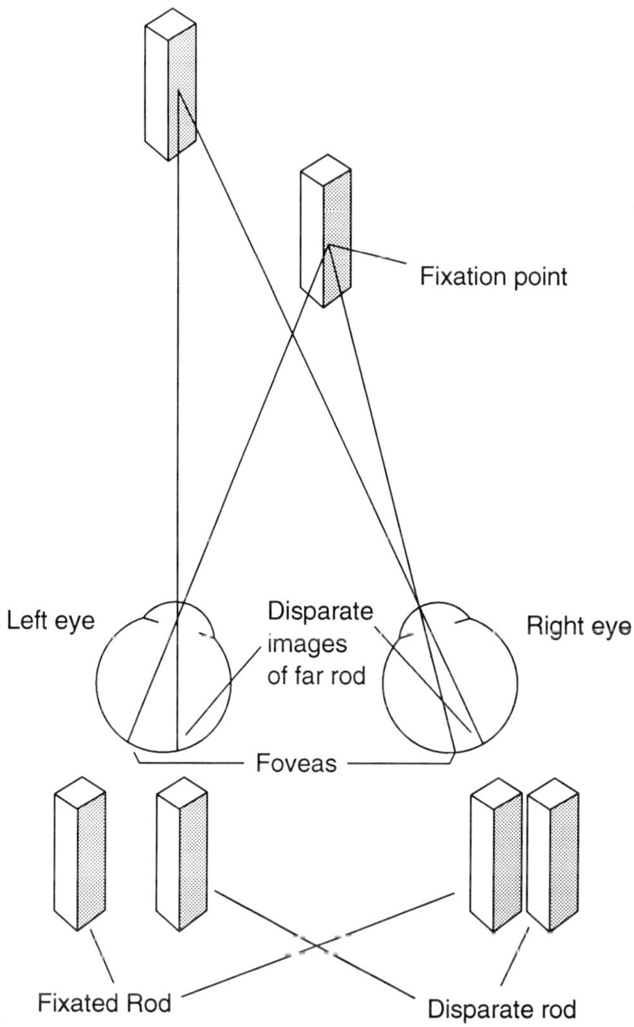

Fig 4.1. The human visual system—the two eyes are fixated on the nearer rod; the more distant rod produces a binocular disparity (a larger distance between the image of the rod and the fovea of the left eye compared to that of the right eye). This effect is illustrated by the corresponding pair of rods below each eye (from Frisby, 1979).

A single human eye is essentially a two-dimensional sensor—it can move elevationally (up and down) and azimuthally (side to side) relative to its position in the head. The angular location of any object viewed by a single eye can thus be specified by the two angles describing the ray of light from the object to the eye. This, in itself, is not sufficient to retrieve three-dimensional information; however the information from the right eye also determines another ray that the point must lie on. The combined attitude information from both sensors together with their known configuration in the baseline of the human skull enables the brain to calculate the true three-dimensional location of the point.

The basic requirement for satellite stereo viewing is that the sensor is capable of viewing the same point on the earth from two different angles. In the case of Landsat (either MSS or TM) the orbit ensures that the repeat path is consistent to within a few kilometres. Thus two images from the same Landsat (path,row) taken at different dates will have virtually the same viewing angle and will not be suitable for use as a stereo pair. Certain areas of the world, however, will lie in the overlap region between two adjacent Landsat paths—one of the satellite tracks passing to the east and one to the west. The overlap between adjacent Landsat paths varies with latitude and can be seen in table 4.1.

Table 4.1. Overlap between adjacent Landsat paths (from Hill 1991)

Latitude	Overlap
0°	7.3%
10°	8.7%
20°	12.9%
30°	19.7%
40°	29.0%
50°	40.4%
60°	53.6%
70°	68.3%
80°	83.9%

Images from adjacent paths acquired in this region will view a point from two different angles and can thus be used as a stereo pair. However, in comparison with the somewhat limited acquisition of suitable stereo pairs from Landsat, the SPOT sensor is designed specifically for stereo viewing, and it is possible to acquire stereo pairs in a variety of angular configurations simply by altering the angle of the sensor mirror prior to image acquisition (see figure 4.2 in the plate section between pp. 84 and 85).

The ability to steer each HRV instrument up to 27° off-nadir allows images of a given area to be acquired more frequently than the nominal 26-day repeat cycle; for example seven passes per 26 days at the equator, and eleven passes at latitude 45°. This provides the ability to acquire stereo pairs of SPOT images of the same area from different viewing angles (see figure 4.3 in the plate section between pp. 84 and 85).

Using SPOT stereo pairs it is possible to produce a DEM with height accuracy of better than the equivalent of one pixel, for example one of the first tests of stereoscopic accuracy of SPOT data using analytic plotters (Konecny *et al.* 1987) achieved RMS errors of 6.5 metres in elevation and an experiment in the Aix-en-Provence area of France also using an analytical stereoplotter (Rodriguez *et al.* 1988) achieved accuracies of under 10 metres in plan and 7 metres in elevation, with a best result of 3 metres in elevation (using a base–height ratio of 1).

4.1.2 Along-track stereo geometry

A second approach to the acquisition of stereoscopic images is to acquire two images at different angles along the same satellite track using an instrument capable, for example, of imaging nadir and forward of nadir simultaneously. The advantages of such an approach include:

— use of the same platform/orbital model for each image;
— near simultaneous acquisition, thus minimising effects due to changing atmospheric and terrain characteristics.

The main disadvantage of along-track stereo is the limited view angle that can be achieved without significant atmospheric distortion—although the subsequently smaller base–height ratio may not necessarily preclude the successful generation of DEMs (Mori *et al.* 1988). Availability of suitable high-resolution data is currently limited to the Japanese JERS-1 satellite, launched in 1992, which is capable of acquiring images with a 20 metre ground resolution and a base–height ratio of 0.3. The JERS-1 optical sensor consists of a seven-band nadir-looking sensor with wavelengths in visible, near infra-red and shortwave infra-red. In addition the instrument has one forward-looking sensor in the shortwave infra-red to produce the stereo capability. Two systems which have been proposed featuring high-resolution (5 metre) along-track stereo are the optical mapping instrument (OMI) (Bagot 1991) and the orbital imaging system (OIS) (Light 1990, 1992) although the future of both these systems is at time of writing (1994) somewhat uncertain.

4.1.3 Photogrammetric methods of height extraction

One simple method of deriving height and contour information from a SPOT stereo pair is to reproduce the left and right hand images photographically and use standard photogrammetric techniques, such as the use of an analytic stereoplotter to produce the desired information. Unfortunately the geometry of a SPOT image is somewhat different from that of an aerial photograph and such techniques are not directly applicable (see section 1.4.1). One solution (Trinder and Donnelly 1988) is to divide the image into a number of segments and to assume that a central point perspective is a fair approximation within that segment (that is, the standard photogrammetric collinearity equations are applicable); alternatively a programmable stereoplotter can be used to relate the image coordinate system to the geographic coordinate system in use to drive the stage plates to their correct position for stereo viewing.

The dynamic orbital exterior orientation for a SPOT image is calculated firstly by modelling the sensor geometry, then calibrating by the use of a few ground control points. These are then used as the input to the process controlling the analytical stereoplotter and form the basis of the inverse collinearity equations used to calculate the instantaneous transformation required to position the right and left hand photographs.

Another possible hybrid approach is the digital stereoplotter. In a typical system the right and left hand images are presented on a split screen image display, but control over the photogrammetric process is still effected manually. The advantages of such a system are that it avoids the image degradation when digital data is written to film and allows contour and digital elevation data to be generated, edited and output in an entirely digital format. A digital stereoplotter uses similar software to calculate the image acquisition geometry and the resultant exterior orientation but instead of mechanically positioning the two SPOT photographs the digital images can be repositioned graphically. The operator views both screens through a mirror stereoscope device and controls the motion via an interactive input device.

4.2 DIGITAL STEREO MATCHING

We have seen how, with a certain amount of ingenuity, terrain height information can be derived by standard photogrammetric techniques. It is desirable, however, both in terms of operator involvement and the ultimate accuracy of the product to generate the DEMs entirely by digital means. There are two main aspects to this—the first of these is the imaging geometry, the modelled based solution of which has been described in section 1.4. The second of these problems is image correspondence which is concerned with the location of the same feature in both images and the calculation of the stereo disparity (and hence the terrain elevation) between them. There are two classes of algorithms for solving the correspondence problem; the first class is based on mathematical correlation of image areas, the second is based on featured-based techniques based on those used for computer vision. In addition, various hybrids of the two approaches are possible.

4.2.1 Correlation methods

The correlation approach to stereo matching (Greenfeld 1991) relies on certain assumptions about the problem domain, primarily that:

(1) it is possible to predefine a coarse mapping between the two images (for example by means of seed points);
(2) the radiometric properties of the two images are similar;
(3) there are no unmodelled geometric distortions between the image geometries;
(4) two image patches to be matched have distinctive texture (smoothly responding surfaces such as ice-sheets can prove very difficult to correlate).

The steps necessary for production of DEMs using the correlation approach can be described in broad terms as follows:

(1) acquire a suitable stereo pair (cloud-free, good base–height ratio);
(2) calculate the imaging geometry for the left-hand image;
(3) calculate the imaging geometry for the right-hand image;
(4) for each point determine the stereo disparity;
(5) calculate the height from the stereo disparity;
(6) perform any interpolation necessary;
(7) reformat onto the final DEM grid.

The use of correlation techniques is broadly similar to that used for the relocation of GCPs (see section 1.2.2). However the relocation of GCPs has a number of simplifying aspects, for example the image will be acquired on the satellite's repeat cycle so that a GCP will be viewed with essentially the same geometry. It is then a simple matter to use one chip as a template and move it in column and row pixel increments until the best-fit position is found. The correlation of SPOT stereo pairs is more problematic, for example they are non-epipolar (an epipolar image geometry means that given a point in the first image, the match point must lie on a known line in the second image). With an epipolar geometry the search space can be restricted to one dimension and is the basis of many stereo matching algorithms, such as those used for robot vision. Unfortunately, due to their dynamic nature described previously, SPOT stereo pairs are not acquired in epipolar format and cannot be converted to epipolar format without further geometric rectification; thus any correlation algorithm must use a two-dimensional search space with a resulting increase in computational complexity. The algorithm is further complicated by two additional factors:

(1) radiometric—the radiometric properties of patches to be matched can vary greatly between the pair, in particular off-nadir images can have noticeable non-linear intensity differences;
(2) geometric—off-nadir viewing combined with terrain variations mean that patches will not always have the same size, shape and orientation.

It is therefore apparent that any correlation scheme used to solve the image correspondence problem must use a two-dimensional search space and be capable of modelling a variety of geometric and radiometric distortions. One of the commonly used algorithms (Gruen 1982;1986) is an adaptive least-squares correlation algorithm. Gruen's algorithm is highly accurate (up to 0.05 pixels has been obtained on aerial photography), however it has a limited pull-in range of about two pixels, and thus must be provided with known seed points from which the matched sheet can grow out. The least-squares criterion is a measure of how well the two images match and is based on minimising the sums of squares of the differences between the corresponding grey levels. The adaptive attribute of the algorithm is that a set of parameters can be specified to allow for geometric and radiometric distortions between the two patches. The radiometric constraint models the transformation between the patches as a linear contrast stretch:

$$I_2(x,y) = a \times I_1(x,y) + b \tag{4.1}$$

where

$I_1(x,y) =$ value at (x,y) in the first image,

$I_2(x,y) =$ value at (x,y) in the second image,

$a, b =$ constants.

The geometric constraints are modelled as affine transformations. This allows a wide variety of scale, translation and perspective changes to be modelled, dealing with all the cases where the patches in both images are small enough to be considered as facets of plane surfaces. Note that, to be effective, the two constraints must be chained together—that is, the radiometric constraint can only be used after the point is mapped into its 'true' position by the geometric constraint. The parameters of the model are altered throughout the search space until the best-fit position is found. The large number of parameters in the model make it impracticable to use a large search space, therefore it is essential that each attempt at correlation starts from a good initial estimate of the corresponding patch. This estimate is known as the seed point of the algorithm and can be derived by various methods such as manual definition, coarse matching on a reduced resolution version of the stereo pair or by use of a feature-based matching algorithm (see section 4.2.2). As with any searching algorithm the result of an attempt to establish corresponding points can have one of three outcomes:

success – algorithm converges to the correct match point;

failure – algorithm fails to converge;

failure – algorithm converges but to the wrong point.

In the first case the outcome is the desired one; in the second case the algorithm 'knows' that it has failed and can attempt to find the correct point using a new seed point. The third case is the most problematic because the algorithm 'thinks' that it has converged correctly (Otto and Chau 1989). It is very difficult to counter false convergence, thus it is essential to have good-quality control over the production of a DEM—this is described in section 4.2.4. The stereo matching algorithm can be speeded up by various means. For example (Norvelle 1992) different shape windows can be used to constrain the search space. This uses the results of matches on neighbouring pixel pairs to determine the size and shape of the area that it is worth searching within. This can also produce more accurate correlations and reduce the averaging effect that may occur when a large window is used over areas of varying terrain. The result of the stereo matching phase will be a measure of the stereo disparity throughout a control network. The stereo disparities will represent the height relative to a floating datum; that of zero parallax. They can now be converted to a fixed datum by incorporating the information contained in the sensor geometry model and the earth model. The result of this is a coarse gridded DEM.

It is also possible to use an iterative method to refine the DEM produced, for example the 'iterative orthophoto refinements' method (Norvelle 1992) uses the following algorithm:

(1) reduce full-scale images by a factor of 4;
(2) perform stereo matching on the reduced-scale images;
(3) calculate an approximate DEM from the stereo disparities;
(4) produce orthoimages of the full-scale input images (using DEM);
(5) perform stereo matching on the orthoimage pairs;
(6) update the DEM based on elevation errors by stereo matching;
(7) iterate steps 4 to 6 until the DEM is sufficiently accurate.

This method uses the fact that if the DEM is accurate it can be used to produce a pair of orthoimages which are virtually identical in their geometry. Thus if used as a stereo pair there will be no noticeable stereo disparities. If however there are inaccuracies in the DEM these will be reflected in mismatches in the stereo matching process which can be used to iteratively refine the DEM. It is important that the orthoimages generated at each iteration are based on the original image data—this removes the problem of image degradation due to repeated resampling.

4.2.2 Feature-based methods

The other possible approach to the stereo matching problem is to use feature-based methods, based on techniques of computer vision. The use of computer vision techniques to model 'real world' features has certain potential advantages. Correlation algorithms can be restricted by their underlying radiometric and geometric constraints and can suffer from 'loss of lock' when they fail to converge. In human vision the most important low-level component is generally considered to be the detection of edges, since the boundary between two features will often be accompanied by a significant change in image intensity over a very small distance; and the problem of edge detection is fundamental to computer-based image interpretation also. The 'edge detector' thought to be employed in the human visual system can be approximated by several difference of gaussian (DoG) filters applied to an image (Kauffman and Wood 1987). Conceptually each DoG filter is used to detect edges that occur at a different spatial frequency and thus can be used to detect edges at a wide variety of separations, unlike the laplacian or other kernel-based filters which are often restricted to edges at a 'one pixel' frequency.

Various approaches to feature matching are possible (see figure 4.4 for one example), the key components in such a matching system being (Greenfeld 1991):

— selection of matching primitives (such as points or edges);
— method for extracting primitives (for example the particular edge detection algorithm used);
— the selection of a suitable search/prediction strategy;
— methods for assessment of results;
— quality control and consistency measures.

A typical first step is an image preparation stage which takes the raw digital stereo images and performs the necessary image rectification to produce a corrected vertically registered stereo pair. This is then followed by a boundary extraction phase (Cooper *et al.* 1987) to extract features from the image pair which are invariant to radiometric effects, such as differences in illumination conditions. One approach is to use DoG filters, applied to the entire image, to detect a candidate set of edge locations to sub-pixel accuracy by the positions of zero crossings (it is generally far easier to determine the position where a function crosses an axis than to determine the maximal location of a characteristically flat correlation peak). The edges are subsequently validated to remove effects of sensor noise, scene noise and characteristics of the DoG filters.

Another approach (Brocklebank and Tam 1991) is to use a context-sensitive filter. This compares the result of an edge extraction filter with an adaptive threshold (defined to be the local intensity gradient). In areas of high edge-gradients the correspondingly high threshold means that only the strongest edges are found, whereas in low-gradient areas weak edges can be found (this avoids the problem of having no edges in some regions of the image which may occur with global thresholding). This should result in a more even distribution of edges over the image. Neighbouring edges are then linked to

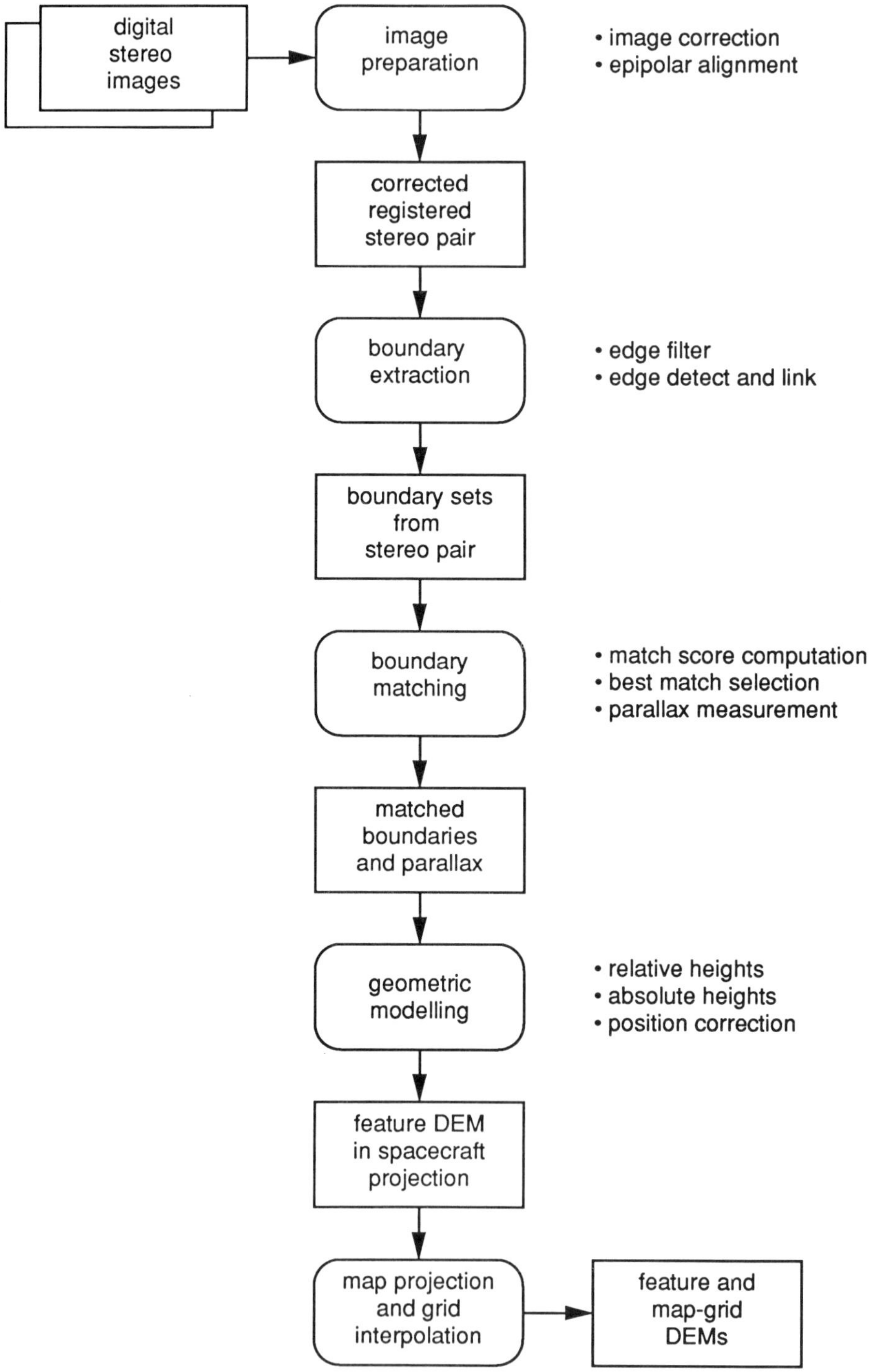

Fig. 4.4. A possible approach to feature-based DEM generation (adapted from Cooper *et al.* 1987)

produce a set of boundary objects. This is performed by a traversal of the candidate boundaries which locates and links the position of zero-crossings, measures their shape, and discards those below significance thresholds for shape and value. The result of this is a series of image features represented in a symbolic vector format which makes the subsequent feature matching process far more rapid than retaining all the original image information. Symbolic representation (Greenfeld 1991) may also be employed at this stage. This involves establishing a list of attributes to characterise a specific edge, such as orientation, strength and relative location. This can be a useful precursor to a boundary matching phase ensuring that only similar edges get considered.

The objective of the boundary matching phase is to determine, for each boundary in the left-hand image, the corresponding boundary in the right-hand image. The algorithm starts with the boundaries produced by the largest DoG filters which will provide the coarsest measure of boundary location. For each feature detected in the first image the best match feature in the second image is determined. Because both features are stored in a symbolic vector format, rather than as raster patches, the matching process is far less computationally intensive. Once the matching boundaries have been detected the stereo disparity between the two boundaries is calculated, typically at separations of one pixel.

Once the initial coarse stereo disparity has been calculated the same process can be repeated at each resolution level as the algorithm focuses in to the disparity at the finest resolution. This hierarchical 'coarse-to-fine' matching system exploits the relationship between adjacent levels to provide a natural description of image features proceeding from the largest attributes down to the finest discernible image detail; this again is a reasonable approximation of how the human visual system operates on a number of hierarchical scales depending on the symbolic nature of the feature being processed. The actual matching criteria are based on a statistical similarity score between the two boundaries being compared; this is followed by the selection of the best such match.

When this has been achieved at the finest level of detail the stereo parallax for each boundary pair is calculated. Thus the result of the boundary matching phase is the set of matched boundaries and the stereo parallax associated with them. Depending on the algorithms used these will be a measure of the stereo disparity at the edges of significant features or alternatively throughout a control network. As with correlation-based methods geometrical modelling can now be used to generate the final DEM—it may also be necessary to use an interpolation algorithm to convert from a feature-based to a grid-based product.

4.2.3 The hybrid approach

Whereas either a correlation or feature-based approach may successfully produce a DEM there are possible disadvantages inherent in each method. For example the correlation method may fail to find sufficient successful matches to produce a DEM, or the feature-based approach may produce good results for identified features, but produce no elevation values at intermediate points. Other problems encountered with correlation-based methods (Kauffman and Wood 1987) include:

— problems with non-linear grey-level effects encountered with off-nadir viewing;
— the effect of different lighting conditions;
— use of algorithms with limited pull-in range, which rely on accurate seeding.

To date there has been a certain reluctance to use hybrid approaches with most practitioners firmly in the correlation or feature-based camps, for example David Marr in his classic book *Vision* asserts that

> '... items will be matchable if they correspond to things in space that have a well-defined physical location. Gray-level pixel values do not. Hence, gray-level correlation fails'. (Marr 1982)

Despite this point of view and the corresponding opinions of the 'correlationists' it would seem beneficial to investigate approaches which make the most of the strengths of each class of algorithm. There are two main types of hybrid processing schemes:

(1) one-step methods where (for instance) a preliminary DEM is produced by feature matching and the fine details are determined by correlation;
(2) adaptive methods where an 'intelligent' controller determines the most appropriate method for a particular situation.

One example of the simple hybrid approach is to use a feature-based method (Barnard and Thompson 1980) to provide the seed points for a Gruen algorithm (Day and Muller 1988). The Barnard and Thompson feature matcher uses an 'interest operator' which can produce results to the order of a pixel accuracy. This is sufficient to be within the pull-in range of the Gruen algorithm which completes the job to sub-pixel accuracy. A similar approach was used for an experimental system to compare the three approaches to stereo matching (Brockelbank and Tam 1991); this system consisted of an area-based module, a feature-based module (using a second directional derivative edge detector and a process of matching stronger features first) and a hybrid module which used the feature-based method to seed the area-based module. Experiments using images of the Dinosaur National Monument (Colorado, USA) and Red Deer (Alberta, Canada) suggested that the hybrid module performed as well as the area-based one, with both outstripping the feature-based approach.

An example of the second method is the approach proposed in a stereo matching system which uses a variety of techniques (Greenfeld 1991). The basic structure of the system can be seen in figure 4.5.

The first phase of the matching is performed by an edge matching process. Subsequently a central monitoring system can invoke one of three additional matching processes:

(1) a correlation-based algorithm;
(2) an 'interest operator';
(3) an interpolator.

This approach

> '... is an enormous challenge in digital photogrammetry' (Greenfeld 1991)

but promises to yield an effective, robust stereo matching system which requires a minimal amount of operator intervention.

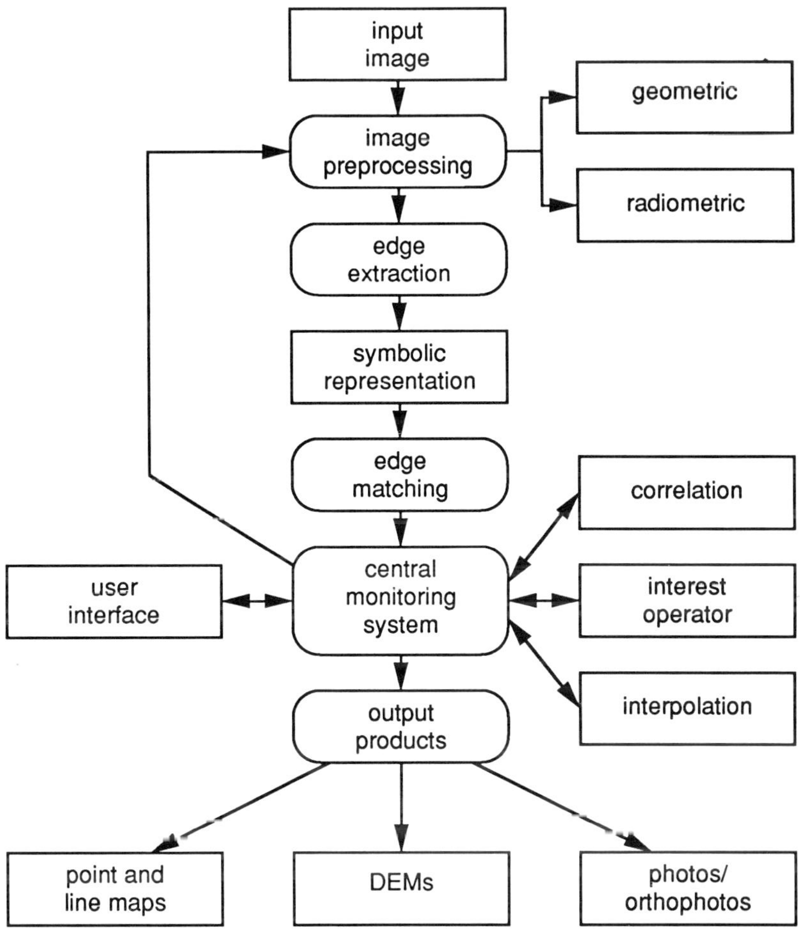

Fig. 4.5. Use of hybrid techniques in a photogrammetric system (from Greenfeld 1991).

4.2.4 Validation

For such a complicated undertaking as the generation of a fine-gridded DEM a substantial degree of quality control is necessary; this can be performed either by visual inspection, or alternatively by algorithmic means. To visually inspect a DEM it can be displayed in either raster or vector format. As a raster image the heights are represented as grey levels, thus the simplest and quickest inspection can be carried out by density slicing. This can be used to reveal unexpected discontinuities in the DEM, discrepancies with known map data, or artifacts caused by attributes of the matching or interpolation algorithms. In vector format the DEM can be presented in many different wire frame or 'blocked in' diagrams. Many graphics processors will enable these to be presented in a variety of different perspectives, and even allow the viewpoint to be changed in real time, for instance to revolve the DEM or to view it from a number of different elevations.

For a more rigorous approach it is necessary to define what exactly is meant by the 'quality' of a DEM, for example (Day and Muller 1988) the following three attributes of quality can be considered:

(1) accuracy with respect to an independently derived DEM;
(2) reliability—defined as the proportion of points with elevation errors greater in magnitude than three standard deviations;
(3) sampling density—defined by the proportion of points actually matched.

Algorithms for automatic quality control rely on applying a number of geophysical constraints to the DEM. For example, if existing elevation data (for example from map contours) is available for the area, constraints such as the minimum and maximum heights for the area can be applied. Other constraints include maximum slope, that is the maximum rate that elevations can change within a prescribed horizontal distance. If the locations of water features can be determined, either from digital map data, or by thematic classification of the images themselves (see chapter 5) then the constraint 'all elevations over a water feature must be equal' can be applied. Thematic information is also important for quality control of extended surface features such as buildings and trees which will introduce additional uncertainty into the elevation determined; for example the satellite will image the top of a canopy of trees rather than the underlying surface topography.

4.3 PRODUCTION OF DEMS FROM SAR

Although the state-of-the-art may not be as advanced as that for optical images, there is great potential in the production of DEMs from spaceborne radar. One of the benefits in using radar is that it measures distances directly, whereas optical sensors measure angles subtended by targets which have to be subsequently converted into distances. The two radar instruments that have potential for DEM production are synthetic aperture radars and radar altimeters. Altimeters have been successfully flown on missions such as Seasat and ERS-1 and are capable of measuring height to within a few centimetres; however due to their large footprint they are not suited for production of local-scale DEMs, thus most research and development to date has concentrated on height extraction from SAR. There are three categories of information inherent in a SAR image (Polidori 1992):

(1) geometric information in terms of range and azimuth;
(2) information from the phase of the radar signal (this is possible because the signal is a complex wave having both real and imaginary parts);
(3) the radiometric information from the backscattered signal.

Each source of information gives rise to a different method of height determination:

— stereo matching (radargrammetry);
— SAR interferometry;
— shape from shading (radar clinometry).

Fig. 2.1. An artist's impression of the Radarsat satellite which will be the first Canadian remote sensing satellite (courtesy Radarsat International).

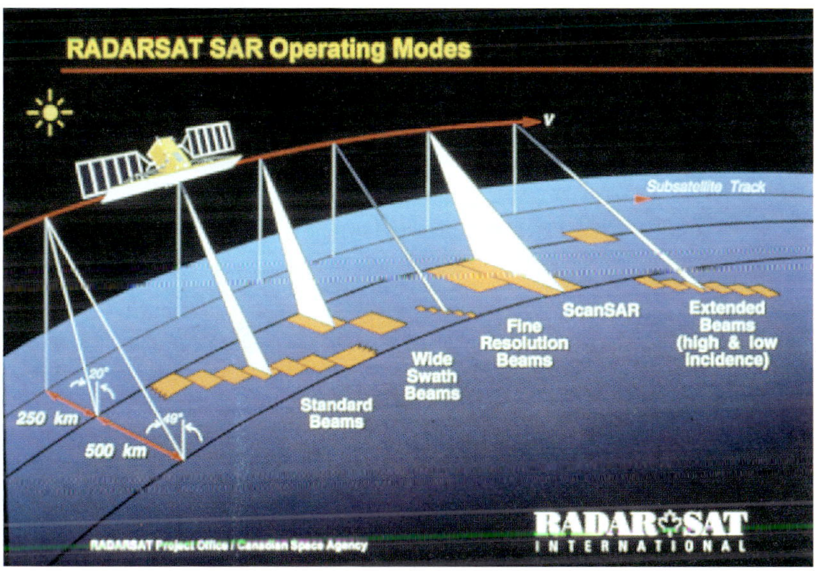

Fig. 2.2. Radarsat operational modes showing selectable viewing angles and variable product sizes (courtesy Radarsat International).

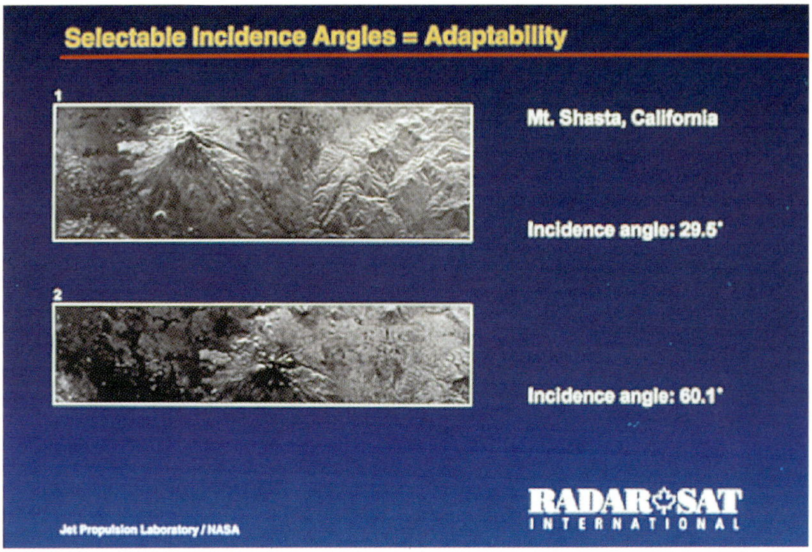

Fig. 2.5. Illustrating the effect of incidence angle on backscatter—the two images of Mount Shasta in California were acquired on consecutive days during the SIR-B mission in 1984. The first image uses an incidence angle of 29.5° and the second 60.1° (courtesty Radarsat International, Jet Propulsion Laboratory/NASA).

Fig 3.8. A mosaic of Europe produced from cloud-free AVHRR data (courtesy DRA Farnborough and NOAA).

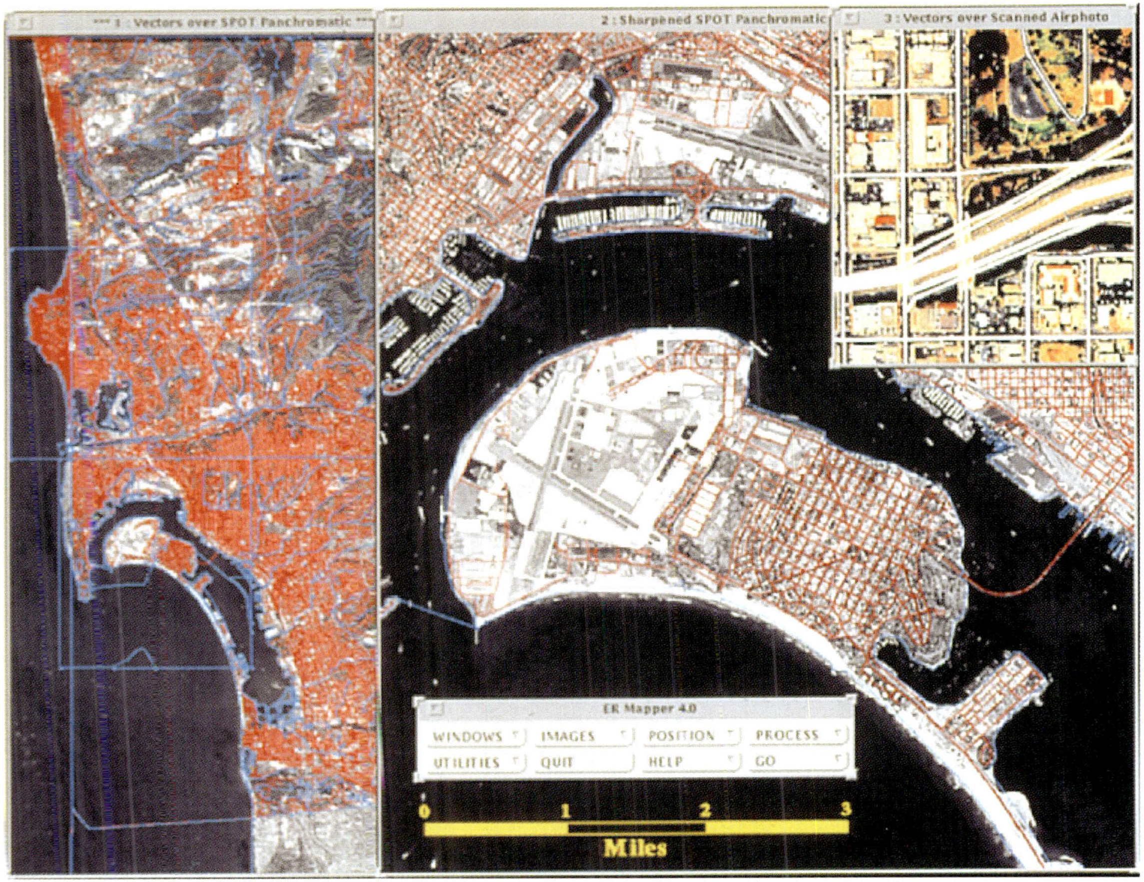

Fig 3.12. Image map of San Diego, California, consisting of digital map data overlaid on geocoded SPOT imagery (courtesy ER Mapper).

Fig. 3.13. Perspective view of Sodaville, Nevada, USA. A Landsat-TM image has been draped over a DEM of the area obtained by image stereo matching (courtesy MacDonald, Dettwiler and Associates).

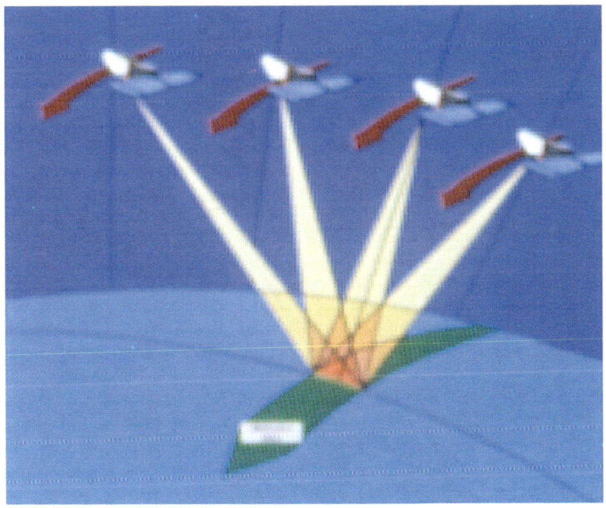

Fig. 4.2. SPOT acquisition angles (courtesy SPOT Image).

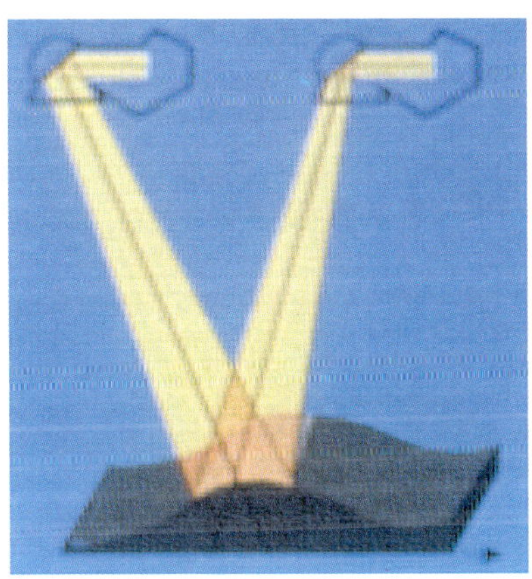

Fig. 4.3. SPOT oblique viewing capability can be used to acquire SPOT stereo pair of a scene from two different orbits (courtesy SPOT Image).

Fig. 5.11. National colour Landsat-TM image of the Adam River area, Vancouver Island, British Columbia, Canada, before classification (courtesy MacDonald Dettwiler and Associates).

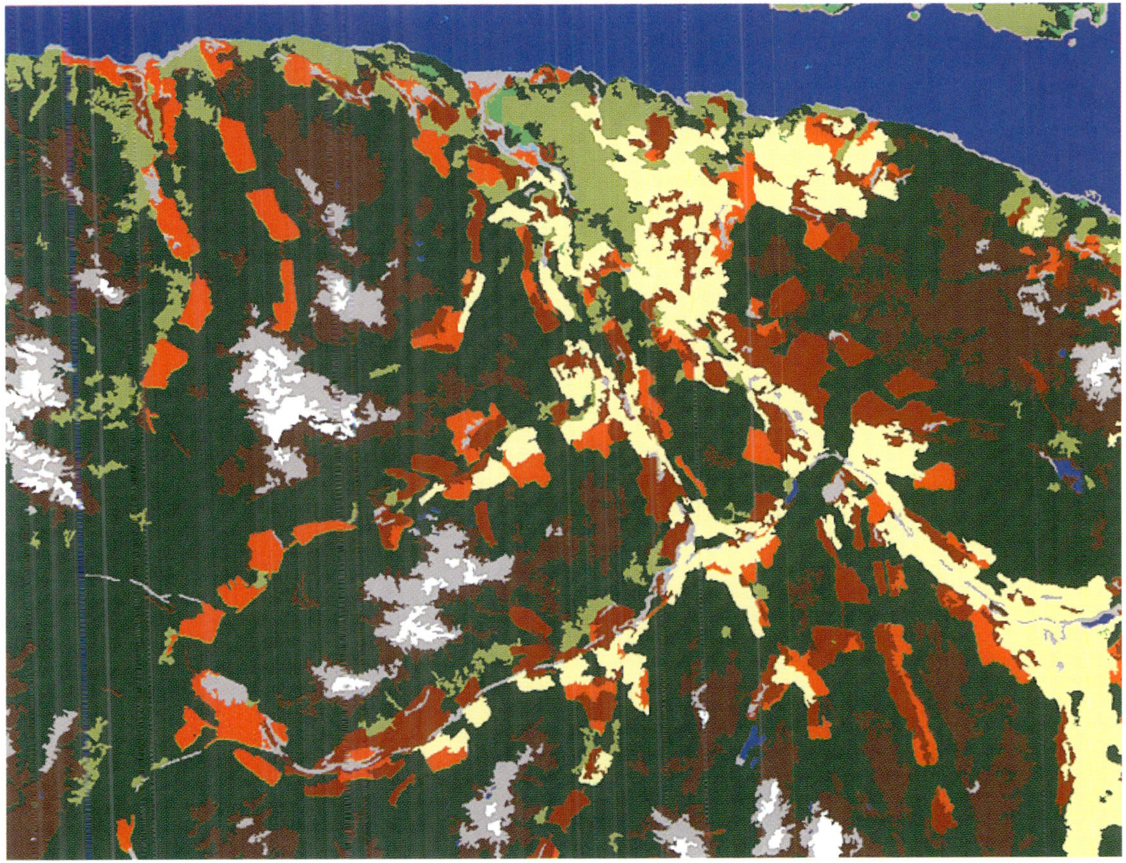

Fig. 5.12. The result of a multispectral classification of the Adam River area shown in figure 5.11. White, grey and brown areas are rock and snow at the summit of mountains, and dark green areas represent original growth coniferous forest. Red shows recent clear-cut areas, with yellow and light-green areas representing past clear-cut areas in various phases of forest regeneration (courtesy MacDonald Dettwiler and Associates).

Fig. 5.18. Multitemporal ERS-1 SAR image of Morecambe Bay in England (see table 5.9 for details of individual images). The pink areas represent the tidal flats and sandbanks exposed at low tide and show a great variety of change compared with the more stable inland areas which appear mainly in shades of grey with some more colourful patches due to the ripening and harvesting of certain agricultural crops (courtesy Radarsat International and European Space Agency).

Operational flowlines for the derivation of height using each of the above methods are shown in figure 4.6.

Derivation of DEMs from SAR is still in its infancy and cannot yet be relied upon to produce consistent DEMs of the accuracy of optical methods. However in some areas of the world it may be the only feasible method. Another attribute of SAR (Polidori 1992) is the synergy between the above methods—radargrammetry derives height values directly at match points whereas interferometry and shape from shading integrate values from pixel to pixel. The relatively crude DEM obtained from stereo matching could be used as a basis to be completed by the other methods, or alternatively to provide constraints to prevent the algorithms 'wandering off' and causing height error propagation.

Another factor to be considered is that the geometry of a particular SAR system may constrain the suitability of a particular approach, for example the incidence angle used by ERS-1 does not allow radargrammetry to be used. It should be noted, however, that a viewing geometry not particularly suited to radargrammetry may well lend itself to other

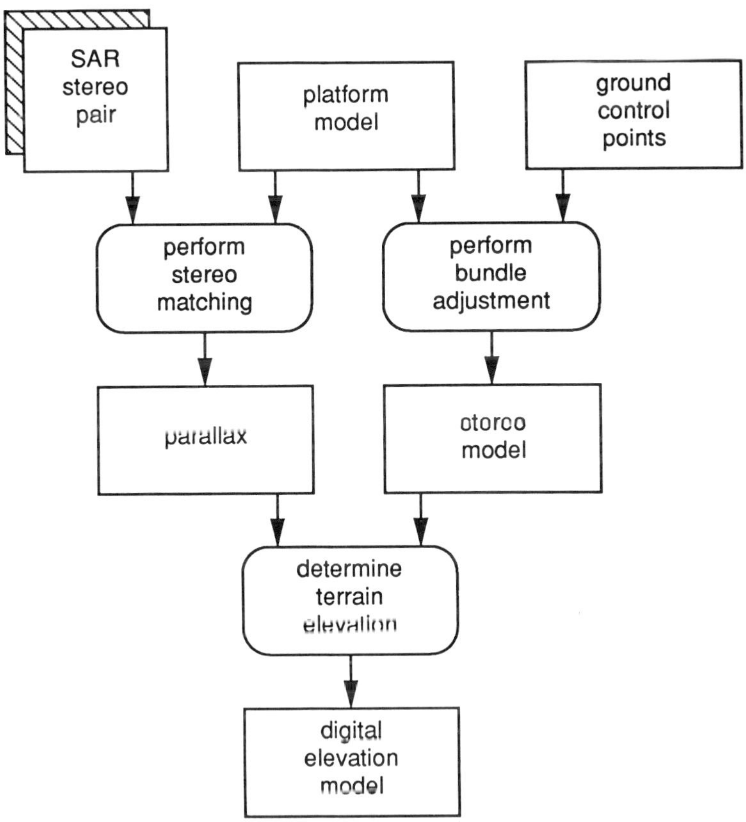

Fig. 4.6. Operational flowlines for DEM generation from SAR (based on Polidori 1992): (a) radargrammetry; (b) interferometry; (c) clinometry.

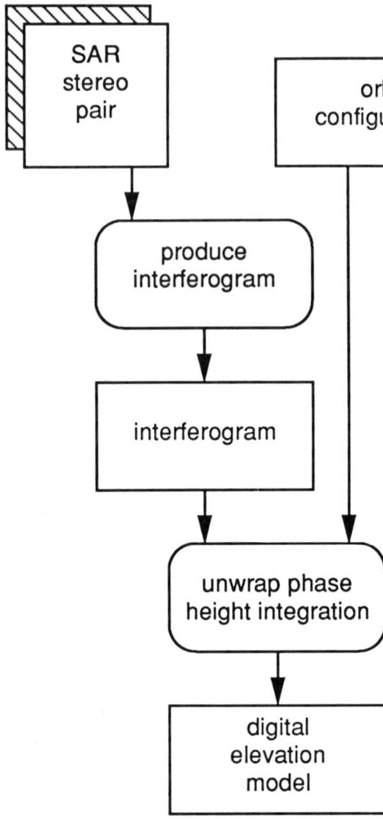

Fig. 4.6(b)

approaches, for example ERS-1 appears to be particularly suited to radar interferometry, thus lending more importance to the synergistic derivation of height from SAR.

4.3.1 Radargrammetry

In common with optical imagery, radar images acquired from two different view angles can be used to form a stereo pair. In order to retain the parallax effect the images are geometrically transformed (Simard *et al.* 1986) to an intermediate projection based on an ideal geoid, that is no correction is made for terrain effects. An additional processing step which is necessary with SAR images is to remove the effect of speckle. This is a random statistical process, due to the coherent nature of the SAR illumination, which causes noise to appear in the SAR image. Most research into stereo matching using SAR concentrates on using correlation techniques rather than feature extraction, as the whole topic of edge/feature extraction from SAR images is not as well advanced as that for optical images. Stereo matching using radar imagery is more difficult than in the optical case (Simard *et al.* 1986) due to the inherent geometric and radiometric characteristics of radar data.

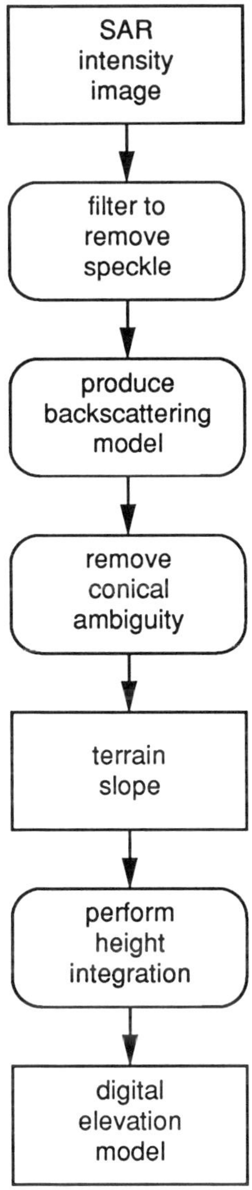

Fig. 4.6.(c).

In general the production of DEMs from stereo SAR imagery using radargrammetry can be accomplished in four phases:

(1) geometric and radiometric preprocessing;
(2) automatic stereo matching;
(3) three-dimensional modelling;
(4) production of the final DEM.

The first stage is to apply a low-pass filter to the input images to remove speckle; however convolution with a low-pass linear filter may have an undesired side-effect by smoothing in the speckle, thus reducing high-frequency information (local height detail) in the resultant DEM. To avoid this, non-linear adaptive filters which remove speckle, whilst preserving high-frequency information, such as the Lee filter (Lee 1981) and the Frost filter (Frost *et al.* 1982) are preferred. There are various possible approaches to the correlation process, for example a method investigated by the Canada Center for Remote Sensing (Simard *et al.* 1986) uses an iterative matching algorithm at different spatial resolutions. The matching of homologous points is performed on a series of stereo pairs with each successive pair being derived by reducing the resolution of the preceding dataset. This acts to constrain the search space algorithm by using the coarse-level locations as starting points for each successive level of the search. This also allows a parallel approach to be adopted for validation as anomalous points can be detected and filtered out at each level of resolution. Interpolation of missing values and low-pass filtering can also be performed at each level of the algorithm.

Stereo reduction is used to optimise the correlation process by using the digital parallax model (DPM) to transform one of the images onto the geometry of the other one. This will reduce the overall parallax and consequently the search space needed for fine tuning. Parallax values can be converted to elevation values using a line-of-sight intersection with the instantaneous platform positions derived from the orbital model. The final stage of processing, as with optical imagery, is to correct the floating height values obtained from the parallax effects into a DEM by comparing them with the reference datum used for the geoid.

4.3.2 SAR interferometry

SAR interferometry represents yet another approach to the derivation of height information from spaceborne sensors. The basic principle is to use two focused complex images gathered by the same SAR sensor in two passes along parallel or crossing orbits. The mathematical definition of a complex signal can be seen in figure 4.7.

Because the signal is complex it can be represented as a 'real' component and an 'imaginary' component (or quadrature). It is usually more convenient to convert these values to a magnitude—phase representation. The phase component of a SAR image (Rocca and Prati 1992) is the result of two mechanisms:

(1) basic phase, proportional to the length of the two-directional path of the radar signal trip;
(2) a phase shift introduced by scattering at the target.

Thus if there is no significant change in terrain scattering effect (for example by a change of surface cover) or elevation (for example due to snow or vegetation changes) between two SAR surveys of the same target from two different locations then there will be a direct relationship between the terrain height and the phase difference between the two signals. The main problem with a phase difference approach (Lin *et al.* 1992) is that the phase difference calculated is the principal value of the phase, that is a value in the range $[-\pi, \pi]$. An infinite number of phase angles can have the same principle value, the relationship being given by the formula below:

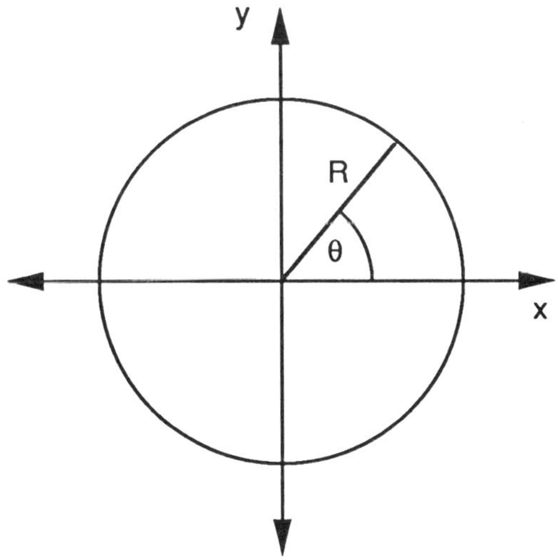

Fig. 4.7. Representation of a complex signal (x + iy) in polar coordinates (R, θ).

$$\phi - \phi_p + 2n\pi \quad (4.2)$$

where

ϕ_p = principal angle in the range $[-\pi, \pi]$,

ϕ = actual angle,

n = any integer.

Thus one of the fundamental tasks of SAR interferometry is to be able to 'unwrap' this phase ambiguity onto its actual value and hence onto the height value. If the phase unwrapping is not performed successfully, local errors in calculating the phase difference will tend to propagate causing global errors in the derived DEM. The basic conditions to allow successful phase unwrapping are:

(1) the magnitude must be non-zero everywhere (a zero magnitude means that the phase is undefined);
(2) the difference in phase between two sample points must be in the range $[-\pi, \pi]$ otherwise there will be an ambiguity in the unwrapping algorithm.

To achieve this it is necessary to account for both phase noise (speckle) and the effects of the phase aliasing. The phase aliasing effect is equivalent to several different height-differences being mapped onto the same phase difference; for instance, in areas of rapidly varying terrain it becomes problematic to determine the true height-difference. In particular pixels with low backscatter are ignored. The low values correspond to a return with a low signal-to-noise ratio—with the noise destroying phase information. One possible approach to removing phase ambiguity is to use a low-resolution DEM to produce seed information about local topography and likely height differences.

The steps in producing a DEM from interferometric SAR are as follows:

(1) SAR processing—it is necessary to start with a pair of undistorted, coherent images free of phase discontinuities; to maintain coherence it is necessary to generate both images using the same doppler centre frequency;
(2) registration—the images are registered using standard correlation techniques to find the best-fit location;
(3) generate interferogram—the phase differences are calculated using the complex form of each SAR image and multiplying the two images using the complex conjugate (the complex conjugate of $a + bi$ being $a - bi$);
(4) remove phase ambiguities—this is the most important stage as it converts the phase differences (which are only known modulo 2π) to actual phase differences and hence to height;
(5) height derivation—using an accurate imaging model the height differences are converted to absolute heights and mapped on to a grid in the desired map projection.

Another approach to the production of interferometric pairs is by means of dual antenna interferometry. This utilises two antennas mounted on the same platform. One antenna produces the radar pulse which is received by both the antennas. This fixed baseline and stable geometry makes it easier to determine phase differences and subsequently heights. Such a method has proved viable in airborne systems and may well be flown on future satellite missions.

4.3.3 Radar clinometry

The human visual system uses a variety of other 'algorithms' apart from stereo information processing, these include:

motion — how the object being viewed changes its angular location with time;
occlusion — whether the object is behind or in front of other objects in the visual field;
shading — variations in illumination response caused by orientation of the object with respect to the light source;
context — acquired knowledge about the scene being viewed.

In a similar way secondary cues could be used to derive three-dimensional information from a satellite image. However a typical image tends not to be as rich in cues as a normal scene viewed by the brain (for example short timescales will generally preclude analysis of motion), and the algorithms needed to determine the information are not fully understood. Nevertheless it is possible to obtain some information about an image from the supplementary information present. For example in optical images an indication of terrain relief comes from the shadow cast from solar illumination. Depending on the sun-elevation angle areas of pronounced relief will have variations in spectral response depending on whether a point is on the bright side or the shadow side of a raised feature. Other information will come from the texture of a region with low-lying areas being characteristically homogeneous, and more mountainous areas showing a more ragged 'crinkly' appearance.

As with optical images it may be possible to obtain information about the topography of a region from the effect it has on an image's appearance—however in the case of SAR images there will be no solar illumination or shadowing effect. Radar clinometry, also known as 'shape from shading' (Guindon 1989) is an attempt to derive three-dimensional information from a single radar image by relating the local terrain slope to the radar backscatter cross-section (Polidori 1992). This is based on a mathematical model of the backscattering process together with certain constraints on the geometry of the surface being imaged. The result of radar clinometry (Thomas *et al.* 1991) is a measurement of slope (local differences in height) which must then be referred to some local datum in order to produce the absolute height values required for a DEM. Research so far has not succeeded in obtaining particular accuracy from this method because of potential sources of error including:

— calibration errors;
— effects of speckle;
— inaccurate model for backscatter;
— propagation of errors/problems converting to height values.

However if DEMs can be obtained in this way they do not require the acquisition of a stereo pair and, because they are likely to be accurate in their high frequencies, there may be a role for radar clinometry, either in combination with other methods or for validation purposes (such as checking consistency with DEMs obtained otherwise).

4.4 APPLICATIONS OF DEMS

4.4.1 Image applications of DEMs

This final section aims to demonstrate the wide-ranging applications of DEMs both synergistically in the production and interpretation of geocoded images (such as their use within the precision geocoding model described previously) and also how DEMs derived from satellite imagery can be used for a wide variety of non-image applications. Table 4.2 gives a cross-reference to the other chapters of the book where DEMs are used in conjunction with image data; a number of other examples are given in Chapter 6 where the role of DEMs within geographic information systems (GIS) is described.

4.4.2 Other DEM applications

The digital format of DEMs make them very amenable to manipulation to derive further information from them. An obvious example is the ability to interrogate a DEM to find the height value at any grid point and, by interpolation, at intermediate points. More complex algorithms can be used to determine the two important secondary indicators of topography (slope and aspect). Another use of DEMs is for intervisibility analysis, that is the determination of which areas can be seen from a given point—applications of this include the siting of communications masts and military planning. Related to this is the calculation of shadow zones which display areas of a landscape that would be in shadow given a specified sun azimuth and elevation. Other DEM manipulations include display of cross-sections across a terrain, the appearance of the horizon from a particular vantage point, and determination of slope values along a road using non-linear profiles. DEMs

Table 4.2. Cross-reference to DEM image applications

Application	Reference
Ground model for precision optical geocoding	1.3.3, 1.4.3
Ground model for precision SAR geocoding	2.4.1
Generation of synthetic SAR images	2.4.1, 2.5
Determination of shadow and layover maps for SAR	2.4.3
Three-dimensional visualisation	3.4.2
Simulation	3.4.2
Animation and fly-bys	3.4.2
Radiometric correction of optical images	5.1.1
Removal of topographic effects in SAR	5.3.2

can also be used to correct for the effect of lengths and areas calculated on a digital map which assume that the surface is flat and viewed from above (Williams 1984). In areas of undulating or sloping terrain the true area will be greater than the estimated area. Slope information from a DEM can be used to correct for this using a simple trigonometric adjustment.

The local topography is an important component of many ecological models and DEMs can be used for modelling catchment areas and watersheds and providing simulation of the flow of water over the earth's surface, which can then be incorporated in hydrological models representing run-off calculations or flood prediction. A DEM will often be an important component of an environmental model within a GIS, for example in soil science, pollution studies or forestry (see section 6.2.3 for details). DEMs can also be used to calculate the volume of material needed to fill in a terrain feature and similar approaches can be used for excavation purposes, either to estimate the volume extracted or to provide a revised model of what the surface would look like, for example, following a mining operation. Other civil engineering applications include road-building/modification and the construction of reservoirs and dams. Finally, visualisation techniques, similar to those described in section 3.4.2, can also use DEMs, for example within detailed flight simulators (Muller *et al.* 1988).

4.5 REFERENCES

Bagot K H (1991) The Optical Mapping Instrument (OMI) 42nd Congress of the International Astronautical Federation, Montreal, Canada, October 1991.

Barnard S T and Thompson W B (1980) Disparity analysis of images. *IEEE Transactions on Pattern Analysis and Machine Intelligence*, **PAMI-2** (4) 333–340.

Belward A S and Valenzuela C R (eds.) (1991) *Remote sensing and Geographical Information Systems for resource management in developing countries*. Kluwer Academic Publishers,

Brockelbank D C and Tam A P (1991) Stereo elevation determination techniques for SPOT imagery. *Photogrammetric Engineering and Remote Sensing*, **57** (8) August 1991, 1065–1073.

Cooper P R, Friedman D E and Wood S A (1987) The automatic generation of Digital Terrain Models from satellite images by stereo. *Acta Astronautica*, **15** (3) 171–180.

Day T and Muller J-P A L (1988) Quality assessment of Digital Elevation Models produced by automatic stereomatchers from SPOT image pairs. *Photogrammetric Record*, **72** (12) 797–808.

Frisby J P (1979) *Seeing—Illusion, Brain and Mind*, Oxford University Press.

Frost V S, Styles J A, Shanmugam K S and Holzman J C (1982) A model for radar images and its application to adaptive digital filtering of multiplicative noise. *IEEE Transactions on Pattern Analysis and Machine Intelligence*, **PAMI-4**, 157–166.

Greenfeld J S (1991) An operator-based matching system. *Photogrammetric Engineering and Remote Sensing*, **57** (8) 1049–1055.

Gruen A W (1982) Adaptive least squares correlation: a powerful image matching technique. *South African Journal of Photogrammetry, Remote Sensing and Cartography*, **14** (3) 175–187.

Gruen A W (1986) High precision image matching for Digital Terrain Model generation. *International Archives of Photogrammetry and Remote Sensing*, **26** (3/1) 284–296.

Guindon B (1989) Development of a shape-from-shading technique for the extraction of topographic models from individual spaceborne SAR images. *Proceedings of IGARSS '89 Symposium*, Vancouver, Canada, 10–14 July 1989, 597–602.

Hill J (1991) Remote sensing systems: sensors and platforms. In Belward and Valenzuela (1991), 111–143.

Kauffman D S and Wood S A (1987) Digital Elevation Model extraction from stereo satellite images. *Proceedings of IGARSS '87 symposium*, Ann Arbor, 18–21 May 1987, 349–352.

Konecny G, Lohmann P, Engel H and Kruck E (1987) Evaluation of SPOT imagery on analytic photogrammetric instruments. *Photogrammetric Engineering and Remote Sensing*, **53** (9) 1223–1230.

Lee J G (1981) Refined filtering of image noise using local statistics. *Computer Graphics and Image Processing*, **15**, 380–389.

Light D L (1990) Characteristics of remote sensors for mapping and earth science applications. *Photogrammetric Engineering and Remote Sensing*, **56** (12) 1613–1623.

Light D L (1992) The orbital imaging system and its potential for Digital Terrain Models of the earth. *Proceedings of Satellite Symposia 1 and 2: Navigation and Mobile Communications, and Image Processing, GIS and Space-assisted Mapping, from the 'International Space Year' Conference*, Munich, Germany, 30 March–4 April 1992, ESA ISY 2, 231–241.

Lin Q, Vesecky J F, Zebker H A (1992) New approaches in interferometric SAR data processing. *IEEE Transactions on Geoscience and Remote Sensing*, **30** (3) 560–567.

Marr D (1982) *Vision*. W H Freeman and Company, San Francisco, California.

Mori N, Takaoka H, Tonoike K, Komai J, Murai S (1988) Investigation of the effectiveness and applications of Japanese ERS-1 stereoscopic images. *International Archives of Photogrammetry and Remote Sensing*, **27**, Part B1/1, 109–119.

Muller J-P, Day T, Kolbusz J, Dalton M, Richards S and Pearson J C (1988) Visualization of topographic data using video animation. *International Archives of Photogrammetry and Remote Sensing*, **27** (B4) 280–288.

Norvelle F R (1992) Window shaping and DEM corrections. *Photogrammetric Engineering and Remote Sensing*, **58** (1) 111–115.

Otto G P and Chau T K W (1989) 'Region-growing' algorithm for matching of terrain images. *Image and Vision Computing*, **7** (2) 84–94.

Polidori L (1992) Potentialities and limitations of topographic mapping from Synthetic Aperture Radar. *Proceedings of Satellite Symposia 1 and 2: Navigation and Mobile Communications, and Image Processing, GIS and Space-assisted Mapping, from the 'International Space Year' Conference*, Munich, Germany, 30 March–4 April 1992, ESA ISY-2, 225–228.

Rocca F and Prati C (1992) Innovative applications of repeated satellite SAR surveys. *Proceedings of Satellite Symposia 1 and 2 : Navigation and Mobile Communications, and Image Processing, GIS and Space-assisted Mapping, from the 'International Space Year' Conference*, Munich, Germany, 30 March–4 April 1992, ESA ISY-2, 193–197.

Rodriguez V, Gigord P, de Gaujac A C and Munier P (1988) Evaluation of the stereoscopic accuracy of the SPOT satellite. *Photogrammetric Engineering and Remote Sensing*, **54** (2) 217–221.

Simard R, Plourde F and Toutin T (1986) Digital elevation modeling with stereo SIR-B image data. *Symposium on Remote Sensing for Resources Development and Environmental Management*, Enschede, Netherlands, August 1986, 161–166.

Thomas J, Kober W and Leberl F (1991) Multiple image SAR shape-from-shading. *Photogrammetric Engineering and Remote Sensing*, **57** (1) 51–59.

Trinder J C and Donnelly B E (1988) SPOT software for Wild Aviolyt BC2 analytical plotter. *International Archives of Photogrammetry and Remote Sensing*, **27** (B4) 412–421.

Williams J M (1984) *REGIS—an RAE experimental image-based Geographic Information System*, Royal Aircraft Establishment (Farnborough, UK), Technical Report 84103.

5

Thematic information from space

As it becomes increasingly important to provide accurate and detailed inventories of surface cover types so it becomes correspondingly difficult to provide all the information required by conventional means. Thematic information derived from geocoded satellite images is becoming a major primary source and in some circumstances may provide the only available solution to the problem. This chapter describes the extraction of thematic information from satellite images. It contains the following sections:

— multispectral image classification: basic techniques for extracting information from multispectral images including the calibration and preprocessing necessary to interpret 'raw' pixel data, the theoretical basis of classification and the two main approaches to classification (supervised and unsupervised) and models for validating the classification results;
— information extraction algorithms: other algorithms for information extraction are described namely principal components analysis (for decorrelating image information), vegetation indexes using visible/infra-red images and extraction of linear and other regular features;
— thematic information from SAR: the basic sources of information in a SAR image, calibration and preprocessing necessary, and methods of classification;
— change detection: the use of image data of the same area acquired at different times for enhanced information extraction via the use of image differencing and other change detection algorithms;
— applications: a few of the many possible applications of geocoded satellite imagery commencing with renewable resources (such as agriculture and forestry), non-renewable resources, and the increasingly important topic of global-scale monitoring and concluding with samples of other applications.

5.1 MULTISPECTRAL IMAGE CLASSIFICATION

5.1.1 Calibration and preprocessing

To appreciate the nature of the information that can be derived from satellite images it is important to understand the physical processes involved and what the digital numbers in an image actually represent. It is assumed that the reader has some knowledge of the physical basis of remote sensing in terms of electromagnetic radiation and its interaction with the earth's surface such as can be found in *Computer Processing of Remotely-Sensed Images* (Mather 1987). In the early development of remote sensing algorithms many researchers manipulated the digital numbers in images directly. This proved satisfactory for qualitative analysis, but for accurate interpretation it is important to be able to calibrate the image values into actual geophysical units. In particular this accurate calibration is an essential prerequisite for the accurate multitemporal analysis described in section 5.4. In a perfect system the image values would correspond directly to the radiance of a point on the earth's surface; however in reality factors such as sensor optics, illumination variation, terrain orientation and atmospheric effects all interact to provide a complicated relationship between recorded and actual values.

A model for calibration can be broken down into the following sub-models:

— illumination model;
— atmospheric model;
— surface interaction model;
— sensor model.

Illumination model

With passive sensors the illumination energy will originate either from the sun (in the case of visible/NIR bands) or from the object itself (in the case of thermal bands). In the case of active instruments such as SAR, or radar altimeter, the energy source is provided by the instrument itself. The maximum spectral radiant energy (Lintz and Simonett 1976) available for an image during daylight hours can be approximated by the spectral distribution of a black body at 5800K (the sun) combined with that of another black body at 300 K (representing the mean temperature of the earth). The two main determinants of the illumination available from the sun will be its distance from the earth (governed by the time of year) and its elevation angle (governed by the time of day).

Atmospheric model

The atmospheric model must account for a number of effects that the earth's atmosphere has on the electromagnetic signal—in the case of passive sensing of a terrestrial target a ray of light from the sun will pass once through the atmosphere on the way to the target, experiencing attenuation effects, then pass through the atmosphere again experiencing further attenuation effects, before finally being detected by the sensor. Depending on the wavelength of the signal, atmospheric gases and aerosol particles will absorb signals, act as scatterers and also as radiant energy sources. One consequence of this is the strong attenuation of electromagnetic radiation in various spectral wavebands, within which the available energy from a transmitted signal has diminished so considerably after passage through the atmosphere as to be virtually unusable for remote sensing purposes. This means that there are only certain atmospheric 'windows', wavebands in which remote sensing is a practical proposition. The major atmospheric windows are shown in table 5.1.

Table 5.1. Main atmospheric windows (based on Lintz and Simonett 1976)

Portion of spectrum	Window (μm)
UV/visible	0.30–0.75
	0.77–0.91
near-IR	1.00–1.12
	1.19–1.34
	1.55–1.75
	2.05–2.40
mid-IR	3.50–4.16
	4.50–5.00
thermal-IR	8.00–9.20
	10.2 –12.4
	17.0 –22.0

The effect of the atmospheric absorption at wavelengths used for earth observation can be seen in figure 5.1.

The two main scattering mechanisms (Chavez 1988) are Rayleigh scattering which occurs when the wavelength of the radiation is much greater than the scattering particles and Mie scattering, which is not as wavelength dependent and is largely caused by dust and smoke particles. Detailed atmospheric models which may be used for atmospheric correction of satellite data, such as 'simulation of the satellite signal in the solar spectrum' (Tanre *et al.* 1990) have been developed. This model (known as 5S) actually consists of eight separate atmospheric models:

— no absorption;
— tropical,
— mid-latitude summer;
— mid-latitude winter;
— sub-arctic summer;
— sub-arctic winter;
— US standard 1962 atmosphere;
— user defined (using radiosonde or other meteorological data).

The last option reflects the fact that, no matter how sophisticated static models of typical atmospheres may be, it is always preferable to model the atmosphere using accurate and contemporary meteorological data.

Surface interaction model

The surface interaction model accounts for one of the major contributions to spectral variation in an image—that provided by terrain variation. The basic imaging geometry is shown in figure 5.2, the three critical variables being the position of the illumination

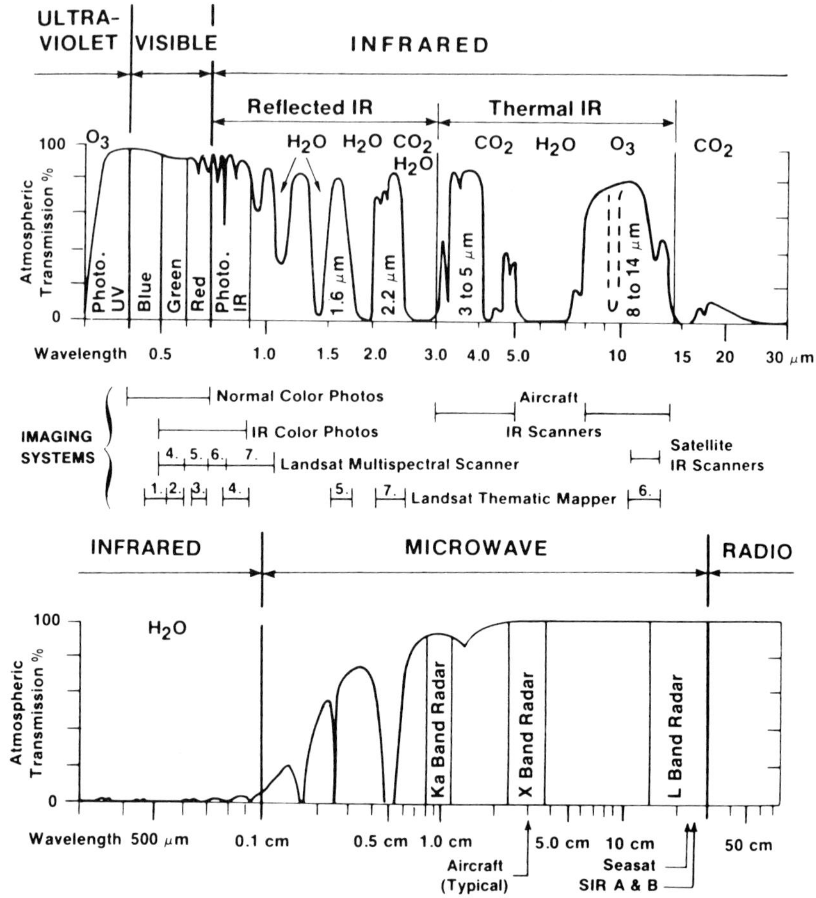

Fig. 5.1. The effect of atmospheric absorption at wavelengths used by various imaging systems (from Sabins 1987).

source (the sun), the target and the sensor. For a given image the satellite orbital plane can be regarded as fixed relative to the sun, thus the major variation is caused by the orientation of the surface facets. Two facets with the same surface attributes but differing slope and aspect will have a different imaging geometry, producing distinct spectral signatures. Thus even in an area with a homogeneous cover, type classification based on spectral signatures may not succeed unless the terrain effect is accounted for.

The simplest form of radiometric correction is to assume that the earth's surface reflects electromagnetic radiation as a Lambertian model. In such a model the only geometrical parameters required are the sun angle (azimuth and elevation), the sensor look angle and the terrain slope and aspect. In general the last two will be unknown for a particular scene. They can however be derived from a numerical analysis of a DEM since the slope represents the rate of change of height at a given point and the aspect represents the direction of greatest slope. Mathematically, slope and aspect can be derived from the

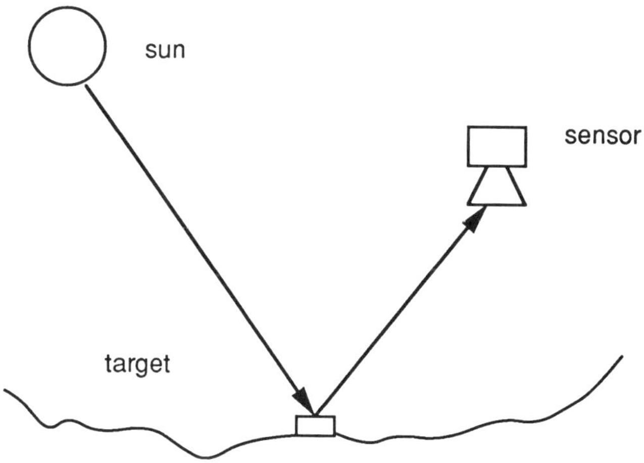

Fig. 5.2. Basic sun–sensor–target imaging geometry.

partial derivatives of a continuous (differentiable) surface. As the DEM is a discontinuous surface (sampled at grid points) it is necessary to estimate the derivatives by numerical interpolation.

Sensor model
The sensor model will be specific to the particular instrument in question (unlike the more generic illumination and atmosphere models). On the other hand, since the sensor is a precisely engineered artifact it is very amenable to mathematical modelling. Typically, pre-launch calibration in a laboratory will give a nominal function to convert from digital numbers to radiance; this calibration will be modified during a sensor's lifetime and coefficients transmitted as part of the spacecraft housekeeping data. Possible methods for in-orbit calibration include:

— ground and atmospheric measurements;
— reference to another sensor;
— a combination of the above methods.

By combining the above four models for a particular image it is possible to calibrate an image more accurately into geophysical units. This in turn provides a potential for a more accurate classification of the image.

5.1.2 Theory of multispectral classification

Spectral signatures
The derivation of feature types based on the multispectral information contained within an image is known as image classification. The colour of an object discriminated by the human eye is determined by the way that it reflects light in the visible spectrum, and it is the eye's sensitivity to light in this fairly narrow waveband that allows it to perceive objects as having different colours. Objects reflect or radiate energy at all wavelengths in

the electromagnetic spectrum, so sensors capable of detecting other spectral bands such as infra-red and ultra-violet will have a greater potential for discrimination of objects than by using visible light alone. Figure 5.3 shows the spectral signatures of three typical land cover classes (bare soil, vegetation and water).

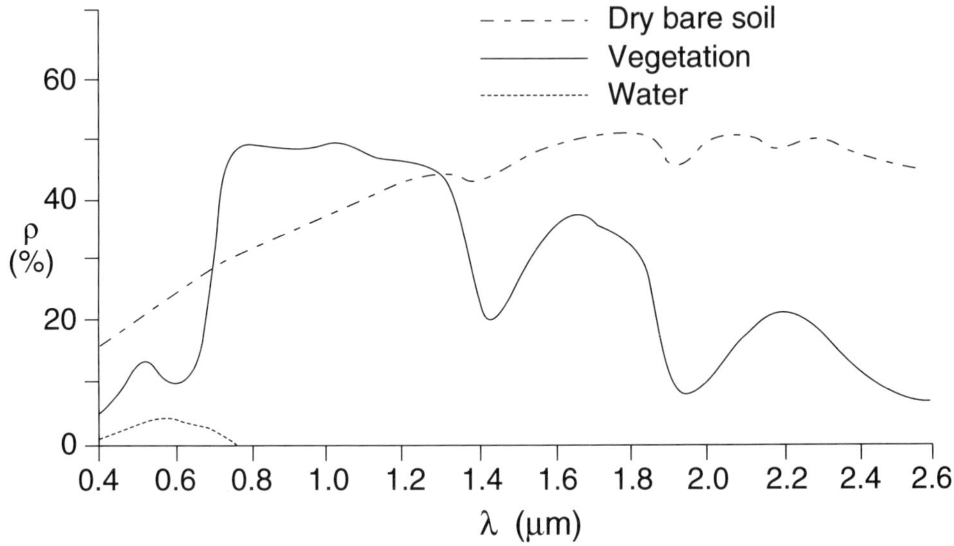

Fig. 5.3. Spectral curves for vegetation, dry bare soil and water (from Bird 1991).

The spectral signature is the typical intensity response of an object as a function of wavelength, and the principle of image classification is to distinguish between different classes on the basis of this signature. Note that the single curves drawn on the diagram are highly schematic; in practice there can be great spectral variation between objects in the same class; thus the simplified curves are better considered as a range of values—an envelope within which spectral values can be expected to lie. To simplify the mathematics of classification various assumptions are generally made, such as:

— the sensor response is uniform across a spectral band;
— each pixel corresponds exactly to a well-defined square on the ground;
— the pixel area is covered with a homogeneous feature type;
— the histogram of each band has a normal (gaussian) distribution.

These assumptions will not be totally applicable in practice, for example a particular sensor will not respond uniformly to all wavelengths within its spectral band, reflectance from neighbouring areas will contribute to each pixel, and one pixel could contain the information from several different classes within its spatial extent—for example it could be part corn-field, part road and part river (the interpretation of such 'mixels' is considered in section 5.1.4).

The assumption about the image histograms having a normal distribution is crucial for the operation of various parametric classification techniques described in section 5.1.4.

Since a typical image histogram will rarely have a normal distribution it may be necessary to preprocess the image by applying a gaussian contrast stretch (see section 3.2.2).

Histogram classification

A simple form of classification of land and sea can be performed by density slicing, that is setting a number of thresholds in an image histogram to divide classes on the basis of their response in a single spectral-band. For example a single threshold can be used to divide the image into two classes based on a simple decision criterion (see figure 5.4):

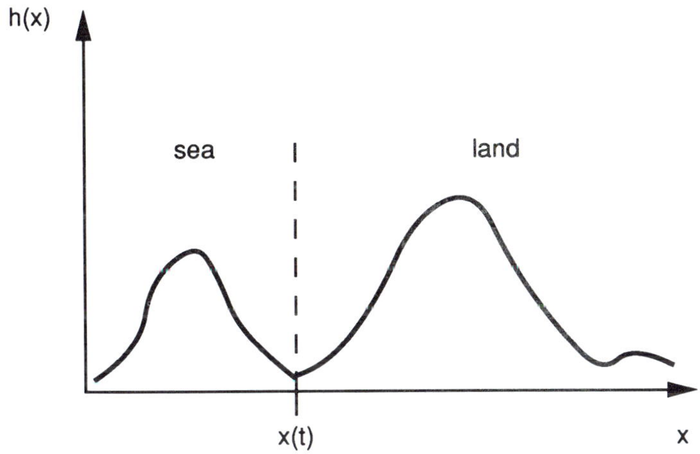

Fig. 5.4. A histogram divided into two image classes by threshold $x(t)$.

— if the pixel value is less than or equal to threshold assign it to class 1 (for example sea);
— if the pixel value is greater than threshold assign it to class 2 (for example land).

This is an example of a binary decision classification. The statistical model used is that the histogram represents the combined effects of two pure populations of pixels each with a characteristic statistical distribution. The classification problem can then be stated in terms of probability 'find the threshold T such that the probability of classifying a pixel as the wrong class is minimised'. If the two histograms do not overlap then any value between the two histogram ranges is suitable and the probability of misclassification is zero. In practice, there will usually be some overlap and thus a possibility that values in the overlap area lead to misclassification (for instance land as sea, or sea as land). In some circumstances improved results may be obtained by using a contour-threading technique (Benny 1980). This makes use of the fact that some interfaces between classes (such as land–sea or river boundaries) can be regarded as a continuous boundary and techniques analogous to those used to thread contours through spot-height data can therefore be used.

BOURNEMOUTH
UNIVERSITY
LIBRARY 70

Classification using scatter diagrams

The probability of misclassification can be reduced by using more than one spectral band and by examining the histograms of the pixel classes in each spectral band. It is most convenient however not to display the image statistic as histograms, but as scatter diagrams. Figure 5.5 shows a typical scatter diagram of two image bands. The x-axis represents all the possible values in band 1 and the y-axis represent the values possible in band 2. The dots on the scatter diagrams are interpreted as indicating combinations of pixel values in the two spectral bands co-incidentally. Thus the dot indicated at (101, 89) indicates that there is at least one pixel in the image with a value of 101 in Band 1 and 89 in Band 2; whereas the blank at (250, 72) indicates that there are no pixels in the image with this particular combination.

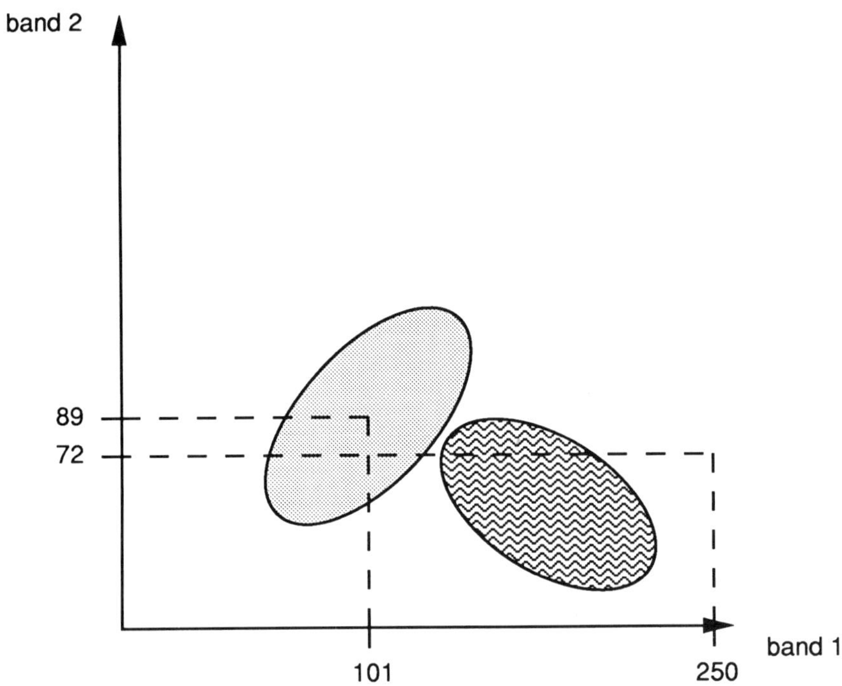

Fig. 5.5. Example of a two-dimensional scatter diagram—image band 2 is plotted against image band 1.

A more sophisticated form of scatter diagram can be produced in the form of a two-dimensional array, I, which is used to count exactly how many co-occurrences of the various pixel values actually occur in the image. Thus, for example, $I(101,89) = 57$ indicates that there are exactly 57 pixels in the image with values of 101 in band 1 and simultaneously 89 in band 2. The potential of multispectral methods can be judged by comparing a two-dimensional scatter diagram with either of the one-dimensional histograms (figure 5.6). In neither case is it possible to separate completely the two classes as

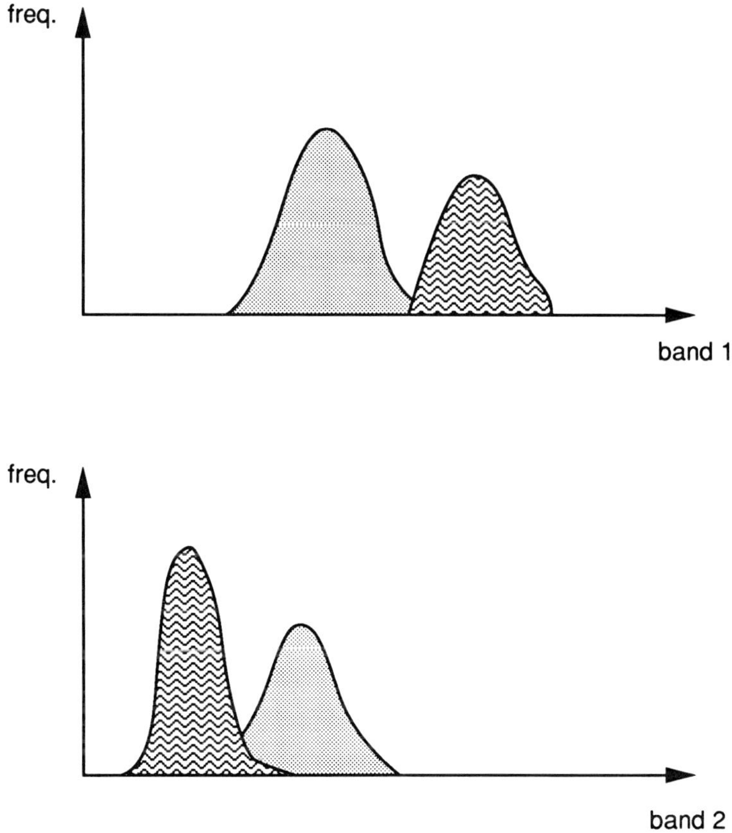

Fig. 5.6. Comparison of the scatter diagram shown in figure 5.5 with the two underlying one-dimensional histograms

there is considerable overlap between them—however, it is a simple matter to separate the classes on the scatter diagram as they form two separate regions (or clusters).

It is thus the task of the image classification algorithm to construct decision boundaries between the clusters to mathematically divide the classes. The methods used can quite easily be generalised to more than two classes but in this case there may again be overlap between clusters. In this circumstance the algorithm must select decision boundaries which minimise the probability of a pixel being misclassified. The methods are easily extended algorithmically to three or more spectral bands but the resulting n-dimensional clusters separated by $(n - 1)$-dimensional decision boundaries cannot easily be visualised.

5.1.3 Unsupervised classification

There are many different approaches to the cluster analysis needed to partition the feature space into distinct classes. Algorithms which use external information (such as incorporation of known classes based on ground data) are known as supervised classifiers, whereas those algorithms which do not, and rely solely on the use of the statistical information inherent in the image, are known as unsupervised classifiers. The aim of an unsupervised classifier is to divide the feature space into clusters which are individually dense yet collectively well separated. In mathematical terms the objective is to minimise the average distance from points within a cluster to the local cluster centroid whilst at the same time maximizing the distance between cluster centroids. A typical algorithm to achieve this is as follows:

(1) choose the number of clusters to divide points into;
(2) for each cluster select an arbitrary starting point to act as centroid;
(3) for each starting point assign every other point to the nearest cluster centre;
(4) for each cluster calculate a new centroid based on points in the cluster;
(5) repeat steps 3 and 4 until the position of the cluster centroids is unchanged;
(6) assign a label to each class.

The labelling of classes is performed by a skilled operator relating the resultant classifications to actual ground-cover types based on 'ground truth' data. In practice (Wilkinson 1991) an implementation of the unsupervised classification algorithm will include extra criteria for refining the classification, for example:

(1) merge rules, which join two clusters together if the distance between their means is less than a specified threshold;
(2) split rules, which break a cluster into two smaller ones if its variance exceeds a specified threshold;
(3) abandoning a cluster and reallocating its members if there are less than a threshold number of them.

Unsupervised classifications are rarely used as the sole means for information extraction from an image, because they use an entirely algorithmic approach, without the introduction of real world knowledge. They can however provide great insight into the feature space representation of an image and provide a good basis for the subsequent use of supervised classification techniques.

5.1.4 Supervised classification

An alternative method (supervised classification) is to provide some information about the pixel classes before the classification takes place. A typical processing flowline for supervised classification can be seen in figure 5.7.

The information about the pixel classes is provided by the operator in a process known as 'training' the classifier. Based on ground information the operator will know which classes some of the areas of the image correspond to. The training algorithm uses the designated areas to extract the sample pixel values in each class. By working out the statistical distribution of the sample values (for example mean, variance) the classification algorithm can then assign class probabilities to all the remaining pixels in the image.

The flowline shown in figure 5.7 also indicates a number of 'feedback loops' used to refine or validate the classification process for example:

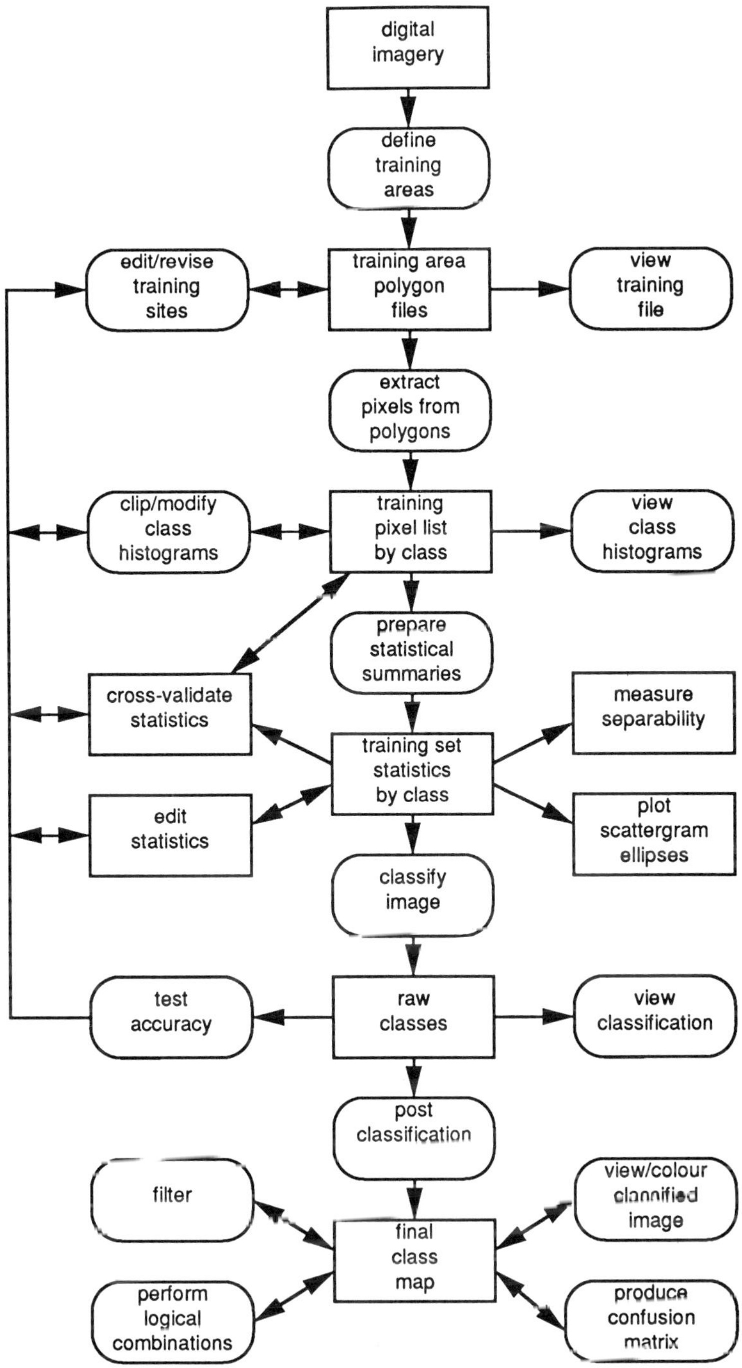

Fig. 5.7. A typical processing flowline for supervised classification (adapted from van Genderen and Uiterwijk 1987).

— edit/revise training sites;
— cross-validate and edit statistics;
— test accuracy;
— post-classification filtering and presentation.

Further details about methods of validation can be found in section 5.1.5.

The box classifier

The simplest supervised classification algorithm is the parallelepiped or box classifier. The idea behind this is shown in figure 5.8.

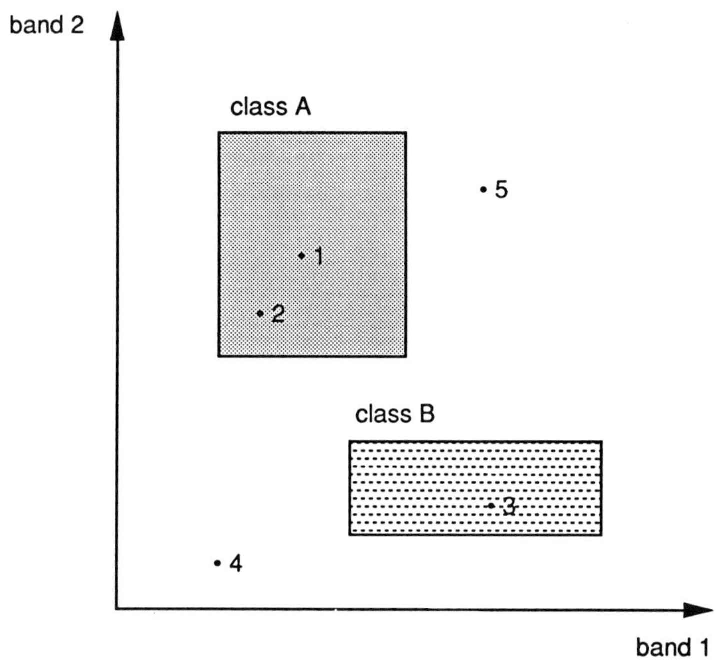

Fig. 5.8. Multispectral classification using a parallelepiped (box) classifier: points 1 and 2 lie in class A, point 3 lies in class B, points 4 and 5 are marked as unclassified.

In two dimensions it may be possible to separate clusters by drawing a rectangle around each one. The training algorithm calculates the size of the rectangle for each sample class; this can be achieved simply by calculating the maximum and minimum values within the sample distribution. A statistically more accurate result is obtained, however, by removing outliers for example using the 5–95 percentile range instead of the entire distribution. The box classification algorithm is 'non-parametric' in the sense that it does not employ statistical parameters to calculate class separation. Its operation is very simple; to classify the entire image it goes through on a pixel-by-pixel basis examining the value in each spectral band used for training. If the pixel value lies within the multidimensional box it is classified as being that feature. In cases where a pixel lies in the overlap between two boxes a decision rule has to be used; a typical rule is to allow the operator to specify which class should have priority.

Parametric classifiers

The box classifier uses a simple decision rule to decide whether a pixel should be included in a particular class or not. There is, however, far more information available in the training samples which can be used to provide better classification estimates. For example a pixel with a spectral value near a class centroid is more likely to be a member of that particular class than one far away from the centroid, or in the overlap of more than one cluster. Classifiers that make use of the statistical distribution of the training samples are known as parametric classifiers. A simple example of a parametric classifier is the 'minimum distance' classifier. This calculates the centroids of each class in feature space (using the training data); the classification then proceeds by assigning each pixel to the class whose centroid is nearest (see figure 5.9).

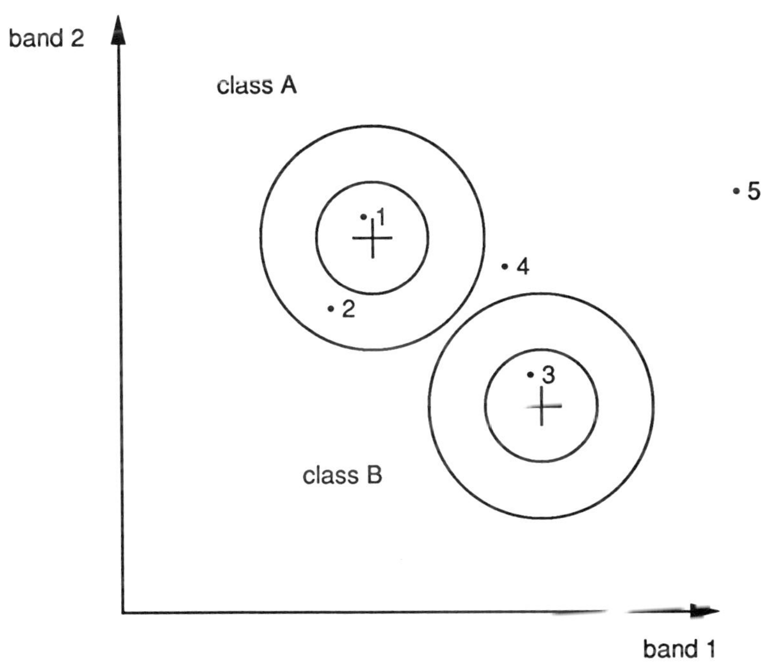

Fig. 5.9. Multispectral classification using a minimum distance classifier: points 1 and 2 lie in class A, point 3 lies in class B, point 4 lies marginally in class A, point 5 lies marginally in class B but with a large probability of misclassification.

This method does have the disadvantage (in a similar way to the box classifier) of assigning a pixel to a class even when it is 'obvious' that there is a greater probability that it belongs to a second class—this is because neither algorithm takes into account the probability density functions that describe the shape of a cluster.

One of the most-used types of parametric classifier is that known as 'maximum likelihood'. This assumes that each class has a multivariate normal distribution; this means that the distance from the cluster centroid, calculated in terms of the variance in each band, can be used as a measure of the probability that the pixel belongs in that class (see figure 5.10). To classify the image each pixel is located in feature space and the maximum likelihood distance to each of the class centroids is calculated. The one given the lowest value is the class to which the pixel is assigned. The distance can also be stored as it gives some indication of the confidence that can be placed in the classification.

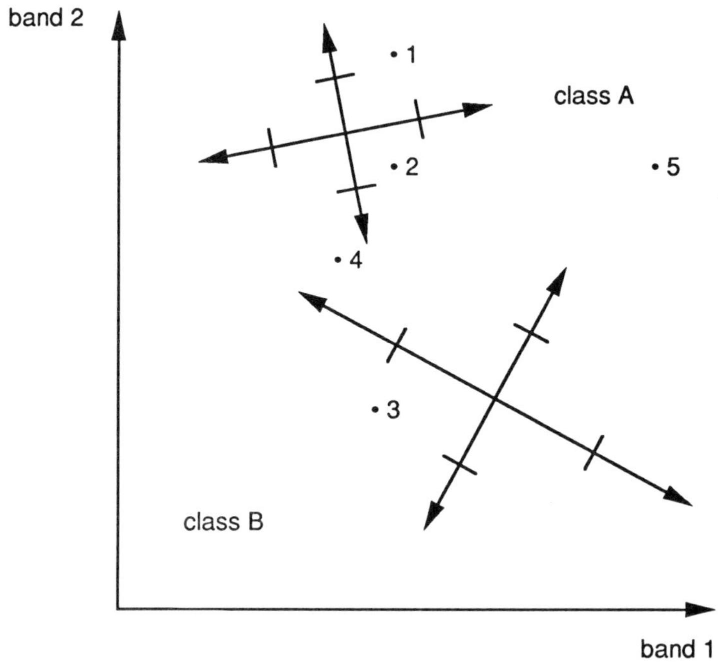

Fig. 5.10. Illustration of a maximum likelihood classification: classes A and B are portrayed by the principal axes of the corresponding feature-space clusters, annotated by 'one unit of distance' tick marks. According to the classification, points 1 and 2 lie in class A, point 3 lies in class B, point 4 lies marginally in class B, point 5 lies marginally in class B but with a large probability of misclassification.

Collection of ground data

For any successful image classification it is important to be able to collect sufficient ground data to accurately calibrate and provide confidence in the subsequent classification. Sources of data can vary from existing maps, aerial photographs and airborne scanner data to field surveys carried out using portable radiometers. The plan for the collection of ground data will be based upon the classes that will be needed for the final image classification. These could vary from a simple single class (for example inland water) through variants in types of woodland and forestry to a complicated hierarchical classification scheme. The choice of an appropriate classification scheme is essential to the successful classification of the image and maximisation of the information derived. A typical hierarchy for classifying land-cover types is shown in table 5.2. This shows the 44 classes (not all of them derivable solely from satellite imagery) used for the Commission of the European Communities CORINE (Coordination of Information on the Environment) project used to define an information system on the state of the environment throughout the European Community.

Table 5.2. Land cover hierarchy (from Cornaert and Maes 1992)

1.	ARTIFICIAL SURFACES	
	1.1 Urban Fabric	
	1.1.1	continuous urban fabric
	1.1.2	discontinuous urban fabric
	1.2 Industrial, commercial	
	1.2.1	industrial or commercial units and transport
	1.2.2	road and rail networks
	1.2.3	port areas
	1.2.4	airports
	1.3 Mine, dump and construction sites	
	1.3.1	mineral extraction sites
	1.3.2	dump sites
	1.3.3	construction sites
	1.4 Artificial vegetated areas	
	1.4.1	green urban areas
	1.4.2	port and leisure facilities
2.	AGRICULTURAL AREAS	
	2.1 Arable Land	
	2.1.1	non-irrigated arable land
	2.1.2	permanently irrigated land
	2.1.3	rice fields
	2.2 Permanent crops	
	2.2.1	vineyards
	2.2.2	fruit trees and berry plantations
	2.2.3	olive groves
	2.3 Pastures	
	2.3.1	pastures
	2.4 Heterogeneous areas	
	2.4.1	annual crops associated with permanent crops
	2.4.2	complex cultivation patterns
	2.4.3	principally agriculture, some natural vegetation
	2.4.4	agro-forestry areas

Table 5.2 (continuned)

3. FOREST AND SEMI-NATURAL AREAS
   - 3.1 Forests
     - 3.1.1 broad-leaved forest
     - 3.1.2 coniferous forest
     - 3.1.3 mixed forest
   - 3.2 Scrub and/or herbaceous vegetation
     - 3.2.1 natural grasslands
     - 3.2.2 moors and heathland associations
     - 3.2.3 sclerophyllous vegetation
     - 3.2.4 transitional woodland-scrub
   - 3.3 Open spaces
     - 3.3.1 beaches, dunes and sands
     - 3.3.2 bare rocks
     - 3.3.3 sparsely vegetated areas
     - 3.3.4 burnt areas
     - 3.3.5 glaciers and perpetual snow

4. WETLANDS
   - 4.1 Inland wetlands
     - 4.1.1 inland marshes
     - 4.1.2 peat bogs
   - 4.2 Maritime wetlands
     - 4.2.1 salt marshes
     - 4.2.2 salines
     - 4.2.3 intertidal flats

5. WATER BODIES
   - 5.1 Inland waters
     - 5.1.1 water courses
     - 5.1.2 water bodies
   - 5.2 Marine waters
     - 5.2.1 coastal lagoons
     - 5.2.2 estuaries
     - 5.2.3 sea and ocean

The result of the data collection exercise will be a database of ground information consisting of a number of training areas together with a feature class assigned from the appropriate hierarchy. These training areas will either be represented in vector format (as a collection of polygons) or in raster format (as a collection of pixels). There are four principal methods of inputting the data required:

(1) direct coordinate input—the ground data is recorded as a list of coordinate pairs (map references) which can be keyed in by hand;
(2) interactive graphic input—the image is displayed on the screen and an operator visually identifies each training area, using a mouse or similar device to draw an appropriate polygon for each training area, or to paint-in its raster extent;
(3) digitiser input—the ground data is presented as overlays on a conventional map-sheet. These can then be digitised with a tablet in the conventional manner;
(4) direct digital input—ground data is increasingly available in digital format; as such data will often be geocoded so it can be directly input into the system.

The specification of training areas is greatly simplified if they are already available in a digital format within a geographic information system (GIS). This is described in more detail in chapter 6.

Operational aspects

Following the input and display of the training data the operator may carry out further addition, modification and deletion of the displayed areas. The first step in the training session is for the operator to specify which bands will be used as the basis for spectral discrimination. A typical IIPF set-up will allow the image bands to be displayed, along with the training areas in some kind of overlay. The pixel values in the overlay are not related to image intensity but are feature codes which indicate the class to which each training area pertains. Based on this collection of information the training session statistics are calculated. The statistics derived from each class will depend on the classification algorithm to be employed. For example in the case of the box-classifier the range and selected percentiles will be calculated; for maximum likelihood the feature centroids and variance vectors will be calculated.

The classification algorithm itself is then applied to the whole image. The result of this will be another image, the same size as the original but with each pixel value being one of the feature codes. This will be similar to the overlay used to display the training areas. It is often beneficial to edit the final output, for example to remove isolated pixels or to correct inaccuracies discovered in the classification. This can either be performed interactively using graphics tools or in some cases automatically. For example an algorithm can be used to remove all groups of pixels less than a certain area (since small groups of pixels are more likely to suffer from spectral contamination making accurate classification more unlikely). If the output classification is intended for a vector-based GIS or digital mapping system the final stage will be the conversion of the classified image from raster format to vector format; this involves taking each connected block of pixels within a class and calculating the polygon that represents the boundary.

An example of an image classification can be seen in figure 5.11 (see the plate section between pp. 84 and 85) which shows a natural colour Landsat TM image of the Adam River area, Vancouver Island, British Columbia, Canada before classification. Figure 5.12 shows the same area following a multispectral classification. White, grey and brown areas are rock and snow at the summit of mountains, and dark green areas represent original growth coniferous forest. Red shows recent clear-cut areas, with yellow and light-green areas representing past clear-cut areas in various phases of forest regeneration.

Mixture modelling

One possible source of error in the classifier can occur if more than one land cover type contributes to a particular pixel—either the pixel is marked as unclassified or is spuriously attributed to the wrong class. The technique known as 'mixture modelling' (Foody and Cox 1991) attempts to 'unmix' a pixel into its components to allow more accurate classifications and areal estimates of cover types. Mixture modelling is based on three assumptions:

(1) surface components are spectrally distinct;
(2) the spectral mixing process can be described by a linear equation;
(3) the only contribution to changes in spectral reflection is that due to the surface components alone.

If the contribution of each component to the mixed pixel is denoted by f_i then two numerical constraints apply (Drake and White 1991):

(1) $f_i > 0$ for all values of i; (5.1)

(2) $f_1 + f_2 + \ldots + f_n = 1$.

The linear mixture model can be solved either by standard least-squares estimation or by using a regularised estimator (Settle 1990) which introduces a prior estimate of pixel mixture throughout the image.

5.1.5 Validation models

The accuracy obtainable in a classification process is dependent on a number of factors. These include the state of the original image (geographic location, illumination, time of year), quality and timeliness of the ground data and whether the image has been geometrically transformed. The measure of classification accuracy will also depend on the classification scheme employed; it is a lot simpler to accurately divide features into two broad classes (such as land/sea) than into thirty to forty different classes separated only by relatively subtle distinguishing attributes.

In terms of the validation models introduced in section 1.3.5, there are two main approaches to validation of the classification process which are generally used in a complementary fashion:

(1) process validation—using the statistics generated by the classification process itself to produce quality control parameters;
(2) product validation—quality control of the final classification product, for example comparing classification output with known land cover categories at sample sites.

The third category of validation, blunder detection, can be difficult to apply in practice as without detailed prior knowledge of the area being classified it is not simple to define what constitutes a 'gross error or physically impossible result'.

Process validation

Process validation consists of the analysis of the statistics generated from the training classes and is thus best suited for use with parametric classifiers. The primary consideration is the 'separability' of the clusters that represent each class in feature space. Separability is based on two parameters:

(1) clusters should be widely separated in feature space (as defined by the distance between cluster centroids);
(2) each individual cluster should be 'compact' (as defined by the variance of the distance of members of a cluster set from its centroid).

The separability of the training clusters can be ascertained interactively by displaying the training data as a scatter diagram. This will indicated 'problem' categories with large variances or which overlap the distributions of other categories. A more rigorous approach is to calculate statistical measures of between-cluster and within-cluster distributions. This will indicate, for every possible combination of pairs of categories, the separability between them.

Product validation

The most common method used for product validation is the confusion matrix (Foody 1992) which is a comparison of the image classes (as determined by multispectral classification) of a sample of points against their actual classes as determined from ground data. A typical confusion matrix can be seen in figure 5.13.

This shows, for example, that 317 sample points were correctly classified as conifer, but that 61 genuine conifer samples were misclassified as deciduous and 23 deciduous samples were classified as conifer. The confusion matrix can be used to measure overall

		Ground Data			
		conifer	decid	agric.	water
Image Classification	conifer	317	23	0	0
	deciduous	61	120	0	0
	agriculture	2	4	60	0
	water	35	29	0	8

Fig. 5.13. Confusion matrix for four classes—conifer, deciduous, agriculture and water (adapted from Foody 1992).

classification accuracy by summing the correctly classified samples (located on the leading diagonal of the matrix) and dividing by the total number of samples. The confusion matrix can also be used to calculate the marginal accuracy attained by a particular class.

Another useful product validation technique is 'residual image analysis' (Jupp and Mayo 1982). This technique involves the construction of a 'mean image' where each pixel is replaced by the mean value of the class to which it has been assigned. The residual image is constructed by subtracting the mean image value from the raw image values on a pixel-wise basis, for each band:

$$R(i,j) = I(i,j) - I^*(i,j) \tag{5.2}$$

where

$R(i,j) =$ value of pixel (i,j) in the residual image;

$I(i,j) =$ value of pixel (i,j) in the original image;

$I^*(i,j) =$ mean value of the spectral class containing (i,j).

The result is an image which displays the variability within each spectral class. Areas of great variability in the residual image will suggest classes for which further selection of training data and possibly refinement of classes may lead to a more accurate overall classification. More advanced 'knowledge-based' techniques for post-classification sorting and validation are described in section 6.1.3.

5.2 INFORMATION EXTRACTION ALGORITHMS

There are a number of other algorithms used to extract information from multispectral images including:

— principal components transform;
— vegetation indexes;
— lineament mapping.

5.2.1 Principal components transform

Statistical analysis of satellite images reveals that there is often a great deal of correlation between image bands; thus, for example, in a scatter diagram plotted of Landsat-TM band 4 value against those of band 5 the characteristic appearance of the cluster will tend to be an elongated ellipse at an angle to the two main axes (see figure 5.14).

The objective of a principal components transform (PCT) is to produce a decorrelated linear combination of the original image bands. In the simplified two-dimensional case this is equivalent to defining two new axes—one along the direction of maximum pixel variability and the second representing variability from this axis. This results in a new 'principal component' containing most of the information (variance) from the original image data and a second component which is not correlated with the first and which contains less information.

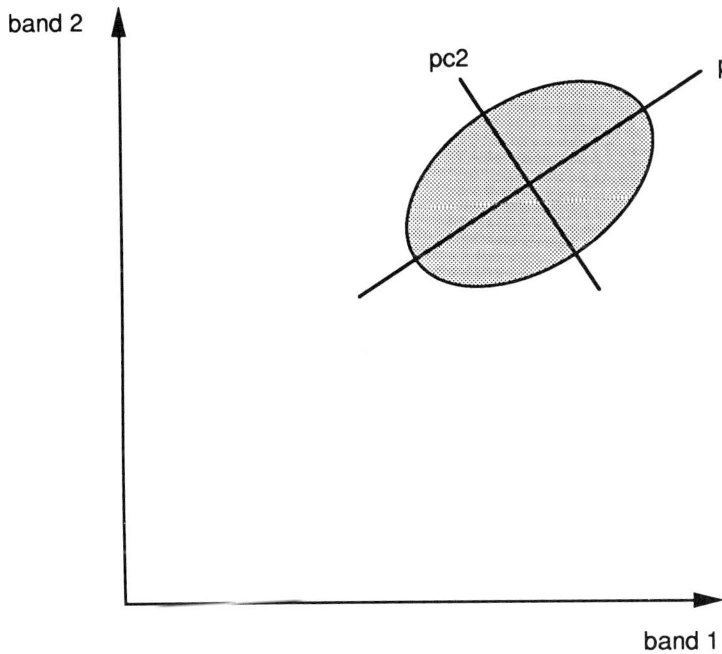

Fig. 5.14. Principal components analysis—two band data is mapped onto principal axes pc1 and pc2.

With more than two image bands the PCT is not simple to visualize but is derived by considering the variance–covariance matrix, *C*(*i*, *j*) where the entries represent the covariance between band *i* and band *j* of the image. For example:

C(2, 3) = covariance between band 2 and band 3;

C(4, 4) = variance of band 4 (equivalent to the covariance between band 4 and itself).

An alternative approach is to utilise the correlation matrix where all entries are normalised by the band variances so that they lie in the range $[-1, 1]$.

The PCT is a linear transformation which, when applied to the image band values, results in a variance–covariance matrix whose non-diagonal entries are zero, this implies no correlation between the bands of the transformed image. In mathematical terms this requires finding the eigenvectors of the matrix *C*(*i*, *j*)—more details of this can be found in specialist works such as *Digital Image Processing* (Pratt 1985).

A similar approach to the PCT is the 'tasselled-cap' transform (Kauth and Thomas 1976, Crist and Cicone 1984) so-called because of its cap-shaped appearance for vegetated areas. Instead of deriving the coefficients of the transformation statistically typical values which correspond to attributes of land cover ('brightness', 'greenness' and 'wetness') are defined. Values of tasselled-cap parameters for Landsat-TM are shown in table 5.3. The following should be noted:

(1) the coefficients define a linear combination of the input bands, for example:

$$\text{brightness} = 0.30 \times (\text{band } 1) + 0.28 \times (\text{band } 2) + \ldots + 0.19 \times (\text{band } 7);$$

(2) the thermal band (band 6) is omitted because of its different spatial resolution;
(3) the (unnamed) fourth, fifth and sixth tasselled-cap bands are usually not used because of their lack of information content.

Table 5.3. Tasselled-cap coefficients (after Wilkinson 1991)

	TM band coefficient					
Name	1	2	3	4	5	7
Brightness	0.30	0.28	0.47	0.56	0.51	0.19
Greenness	–0.28	–0.24	–0.54	0.72	0.08	–0.18
Wetness	0.15	0.20	0.33	0.34	–0.71	–0.46
(four)	–0.82	0.08	0.44	–0.06	0.20	–0.28
(five)	–0.33	0.05	0.11	0.19	–0.44	0.81
(six)	0.11	–0.90	0.41	0.06	–0.03	0.02

5.2.2 Vegetation indexes

Multispectral classification is a general-purpose tool which can be used to classify an entire image into land-cover types. More specialised information extraction algorithms can also be used. One example of these are vegetation indexes which make use of the fact that information in a multispectral image is typically concentrated within a small number of spectral bands, for example (Sheffield 1985) after a PCT 99% of the information in a 7-band Landsat-TM image is usually contained within the first three principal components. Vegetation indexes generally make use of two spectral bands, one in the red part of the spectrum and one in the near-infra-red as these portions of the spectrum are optimal to distinguish vegetation from background soil. For example (Richardson and Wiegand 1977) the perpendicular vegetation index (PVI) defines a characteristic soil line in the two-dimensional feature space defined by a red band (Landsat MSS band 5) and a near-infra-red band (Landsat MSS band 7). The vegetation content of a pixel is then defined in terms of its distance from this line.

Since their original development for Landsat data a number of vegetation indexes have also been successfully used to estimate vegetation from AVHRR data—although originally intended for meteorological purposes such as cloud detection and measuring of sea-surface temperatures, the AVHRR instrument has several interesting properties that lends itself to land cover monitoring including a much wider area coverage than Landsat, a rapid revisit capacity (often daily) and a large pixel size leading to cheaper processing costs per unit area. AVHRR bands provide a suitable basis for vegetation mapping because the red band (band 1) is sensitive to absorption by the chlorophyll in vegetation up to about 0.7 μm. At longer wavelengths lack of absorbing pigments in the green vegetation together with the leaf structure leads to a much higher reflectivity. Soils on the other hand are characteristically more reddish or neutral in colour so spectral analysis based on the ratio of near-infra-red (band 2) to red (band 1) can be used to determine the amount and categories of vegetation cover and subsequently to derive indicators such as leaf-area index (LAI), green-leaf biomass and total biomass. Some typical AVHRR vegetation indexes making use of AVHRR bands 1 and 2 are shown in table 5.4.

The principal application of AVHRR vegetation indexes is described in more detail in section 5.5.3.

Table 5.4. AVHRR vegetation indexes (based on Hayes 1985)

Name	Acronym	Formula
Environmental vegetation index	EVI	NIR – red
Infra-red over red	IR/R	NIR/red
Normalised difference vegetation index	NDVI	(NIR – red)/(NIR + red)
Transformed vegetation index	TVI	$(\text{NDVI} + 0.5)^{1/2}$
Vegetative sponge index	VSI	sponge × NDVI

where

red = band 1 of AVHRR (visible),

NIR = band 2 of AVHRR (near-infra-red),

sponge is an environmental moistness variable.

5.2.3 Lineament mapping

Another important source of information in a satellite image is the location of linear features which can typify roads and communication features, geological lineaments or boundaries between land-cover types. A simple approach to detection of linear features is to use linear convolution filters (of the type described in section 3.2.3) in order to detect the edges in an image (that is pixels that exhibit a marked change in grey-level from their neighbours). Typical filters used are the laplacian kernel (which responds strongly to isolated points) and the series of Sobel filters (which respond strongly to edges in a given direction). The application of a filter to an entire image results in a map of the edge strength at each pixel. By applying a threshold value only the strongest edges are retained resulting in an edge-map of the image. This method does however have a number of drawbacks:

(1) edges may not be detected in areas of low-pixel values because the correspondingly subtle changes in relative value at an edge-pixel are not detected by a linear filter which measures absolute changes in value;
(2) if the threshold is set too high many edge-pixels may be excluded unnecessarily;

(3) if the threshold is set too low spurious edges (due for example to noise in the image) may be inadvertently included.

Solutions to these problems include using non-linear filtering, use of adaptive local thresholds and use of constraint rules to link neighbouring edge candidates together. More details of the use of knowledge-based approaches can be found in section 6.3.1.

Another approach to the detection of spatial features (Cross 1988) is the use of the Hough transform. The Hough transform is concerned with the shape of features rather than their spectral properties and has been used widely in image processing applications (such as robot vision) for the location of features whose general shape (for example line, circle, or 'V') is known but exact size, position and orientation may not be. The application of the Hough transform consists of three phases:

(1) construction of an edge map;
(2) transformation into parameter space;
(3) extraction of features from parameter space.

The edge map is produced by using linear filtering followed by thresholding to produce candidate edges with the strongest response. The parameter space used depends on the type of features it is desired to detect, for example a straight line can be represented by the equation:

$$y = mx + c \tag{5.3}$$

where

$m =$ gradient of the line,

$c =$ intersection on the y-axis.

In this case the two parameters defined are m and c and any line in (x, y) space can be represented as a single point in (m, c) space. The Hough transform works by using an accumulator array in parameter space—for each edge candidate there is a whole family of lines which could pass through it which correspond to a number of bins in the parameter accumulator which are incremented. The bins with the most entries in at the end of the process represent the strongest linear features.

5.3 THEMATIC INFORMATION FROM SAR

5.3.1 Sources of information in a SAR image

The processing of SAR for thematic mapping applications differs fundamentally from that of optical imagery and it is still in a far less understood state of research. Spaceborne SAR images are (generally) single-frequency, thus cannot be classified by multispectral means, and have characteristically different statistical distributions. The mechanism whereby incident microwave radiation interacts with a surface area is complex and imperfectly understood, and the aim of much research is to provide better models of the fundamental imaging process which can subsequently be used as the basis for more advanced classification methods. There are some characteristic relationships between various types of target and their appearance in a SAR image; for example specular reflectors such as lakes and rivers will appear very dark, and good corner reflectors such as bridges as very bright tones. The basic interaction between radar waves and a surface target depends on how rough the target is (Churchill and Sieber 1991). In mathematical terms roughness can be defined by threshold equations such as the Rayleigh criterion which defines a rough surface as one that obeys the equation:

$$\Delta h \cos\theta > \lambda/8 \tag{5.4}$$

where

$\Delta h =$ average vertical displacement of the surface,

$\theta =$ incidence angle of the radar beam,

$\lambda =$ wavelength of the radar beam.

If, according to the above equation, a surface is smooth then mirror-like specular reflection will occur, whereas for rougher surfaces scattering will occur in many different directions. Examples of surface types of varying roughness and the effect on image appearance can be seen in table 5.5.

Table 5.5. Example of surface roughness and consequent image appearance (after Stone 1986)

Roughness	Image appearance	Target description
Very smooth	black	absolutely smooth and flat, open water, fine-gravel mud or silt
Smooth	dark grey	smooth and flat with minor perturbations, sand, no vegetation
Intermediate	mid-grey	gravelly/stony areas, ploughed fields with shallow furrows, low vegetation
Rough	light grey	deeply ploughed fields, small cobbles and boulders, grass and low scrub vegetation
Very rough	white	gullying, boulders, tall widely spaced vegetation, blocky rock outcrops

The dependence of backscatter on radar geometry shows that two of the most important radar parameters for target discrimination are incidence angle and resolution, the values of which for the major SAR missions are shown in table 5.6. The number of looks shown in the table is that required to achieve the azimuth and range resolution indicated.

Figure 5.15 shows the definition of incidence angle which is the complement of the depression angle. The dominant control on backscatter at various incidence angles is shown in figure 5.16; see also figure 2.5 for an illustration of the effect on actual SAR images.

In general (Stone 1986) shallow incidence angles such as that obtainable with SIR-A (47°) are better for surface roughness discrimination; steeper incidence angles such as

Table 5.6. Incidence angle and resolution for SAR sensors (based on Churchill and Sieber 1991)

Sensor	Incidence angle	Resolution		
		Azimuth (m)	Range (m)	Looks
Seasat	20°	25	25	4
SIR-A	47°	40	40	4–7
SIR-B	15–60°	25	17–58	4
ERS-1	23°	30	30	4
JERS-1	35°	18	18	3
ERS-2	23°	30	30	4
SIR-C	15–55°	30	10–60	4
Radarsat	20–50°	4–100	4–100	1–8

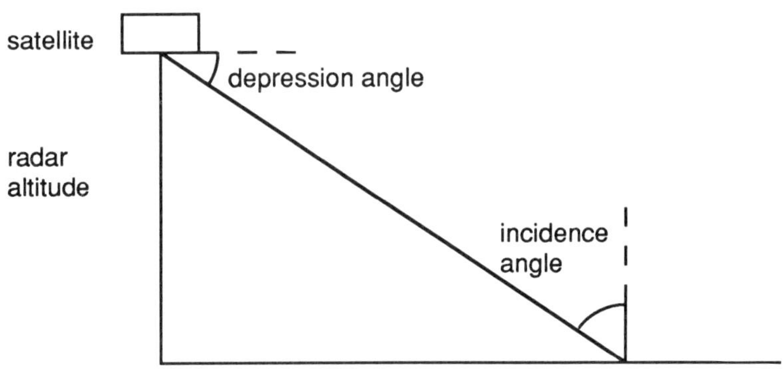

Fig. 5.15. Relationship between depression and incidence angle.

Seasat (20°) and ERS-1 (23°) are more appropriate for areas of high relief. Another important parameter is wavelength (Churchill and Sieber 1991); at radar wavelengths the signal recorded is determined to a great extent by scatterers of characteristic size 1–10 centimetres such as leaves, stems and branches of the target. The main bands used for spaceborne SAR, using the standard IEEE nomenclature (IEEE 1976) are shown in table 5.7.

The last important parameter for SAR imagery is polarisation which will either be horizontal or vertical. Since either type of polarisation may be specified for transmit or receive there are a total of four polarisation options (see table 5.8).

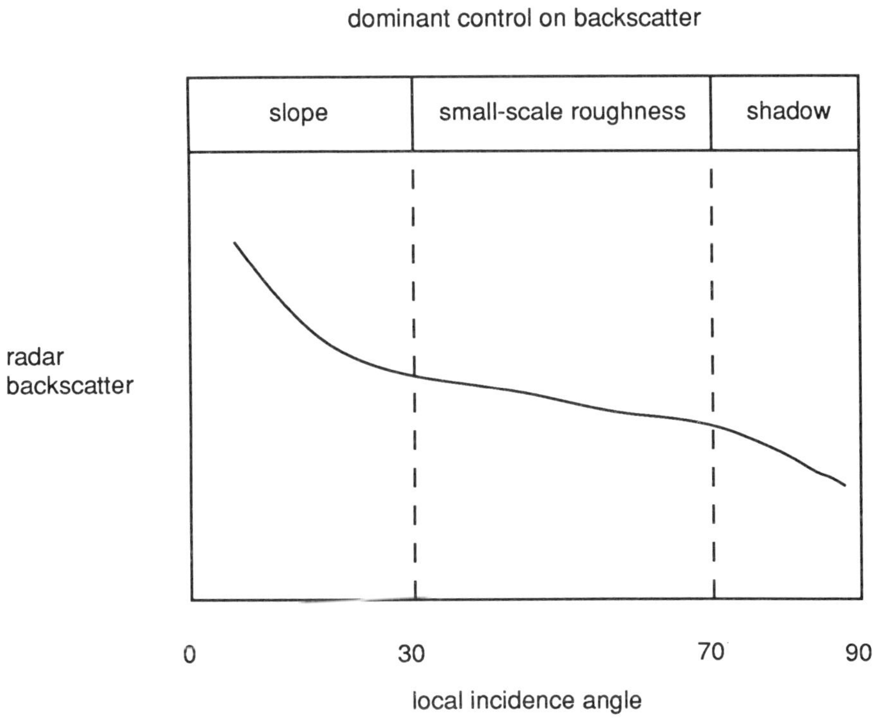

Fig. 5.16. Dominant control on backscatter at various incidence angles; see also figure 2.5 for an illustration of the effect on actual SAR images (based on Stone 1986).

Table 5.7. The main radar bands for spaceborne SAR (adapted from Churchill and Sieber 1991)

Band name	Frequency (GHz)	Wavelength (mm)	Sensors
L	1–2	150–300	Seasat SIR-A SIR-B SIR-C JERS-1
C	4–8	37.5–75	SIR-C ERS-1 ERS-2 Radarsat
X	8–12	25–37.5	SIR-C

Table 5.8. Polarisation options

Designator	Transmit	Receive	Type	Sensors (present and future)
HH	horizontal	horizontal	like	Seasat SIR-A SIR-B JERS-1 Radarsat
VV	vertical	vertical	like	ERS-1 ERS-2 SIR-C (X-band)
VH	vertical	horizontal	cross	SIR-C (polarimetric C and L bands)
HV	horizontal	vertical	cross	SIR-C (polarimetric C and L bands)

5.3.2 Calibration and preprocessing

The two main complications which preclude the use of simple image analysis techniques for thematic analysis of SAR data are:

(1) speckle which arises from coherent, monochrome microwave radiation and results in high variance of image tone resulting in the likely failure of pixel-based methods;
(2) the viewing geometry.

The removal of speckle can be attained by the overlaying of several looks of the same area and hence averaging out the speckle values or by the use of non-linear filters such as median filters or the Lee filter (Lee 1981) to remove speckle without smoothing the genuine image data unduly.

Topography can have a significant effect on image tone. Experiments with airborne SAR (Foody 1986) indicate that variations in topography can account for 5–25% of the variance in image tone. This results in an increased range of possible values for each class of land-cover type leading to overlaps between classes and possible misclassification. The brightness of the target in a SAR image is dependent on the angle between the surface facet and the incoming radiation, hence it is a function both of the incidence angle and the local terrain slope. It is highly desirable that the effect of slope is removed as part of the calibration process before any quantitative classification is made. This is most effectively performed with the use of a digital elevation model. Another useful preprocessing technique (Foody 1987) makes use of fitting a smooth curve to a radiometric profile of the data in order to eliminate local effects.

5.3.3 Classification methods

The classification of SAR images can be divided broadly into four categories (in increasing order of sophistication):

(1) cerebral—visual photo-interpretation of SAR images as photographic products;
(2) image processing—use of simple enhancement techniques to improve the interpretation using an IIPF;
(3) information extraction algorithms—use of statistical and other mathematically based tools to help automate the interpretation process;
(4) model based—use of specific models of SAR geometry, target type and surface interaction.

The simple approach to classification of SAR images is to examine them visually, using a hard copy image, or an IIPF in the manner of a photo-interpreter. The two main objectives for manual interpretation of a feature (Quegan *et al.* 1985) are tone (which can be identified on a subjective scale—very light, light, medium, dark and very dark) and texture (on a similar scale—very smooth, smooth, mid-texture, rough and very rough).

Image processing tools can be of use to remove the effects of speckle and image geometry. Standard image processing techniques such as contrast stretching will also improve interpretability, but the use of linear edge-enhancing filters will often create spurious detail as the result of speckle so are not normally applicable.

The main information extraction algorithms are concerned with the statistics of a SAR image (such as mean, variance and higher-order textural statistics of backscatter) and may be obtained on a per-pixel or per-region basis. Again, the speckle inherent in the images makes image segmentation techniques difficult to apply, and a better result may be obtained using digitised map information such as field-boundaries (Wooding 1985). An alternative approach, and one suited to minimising the effect of image geometry on classification accuracy (Foody 1987), is to divide the image into sectors, either crudely based on equal divisions of ground-range distance or more accurately based on breaks in slope on the radiometric profile.

The model-based approach to SAR interpretation is based on identifying the main parameters of a surface type that will effect the radar backscatter. As described previously, roughness is an important parameter in such a model and so is the target's dielectric constant; for example the moisture content of crops will have a significant effect on the backscattered signal. A detailed model for a particular target type will also need to include the background material and the geometrical relationships between individual scatterers that comprise the target. For example a model used for radar classification of forestry (Simonett *et al.* 1987) has the following major components:

(1) direct backscattering from the soil surface;
(2) backscattering from large branches and trunks;
(3) volume scattering from foliage, canopy and leaf litter;
(4) interaction caused by corner reflection from tree trunks;
(5) other interaction such as multiple scatterings between foliage, ground surface, shrub canopy and branches.

Temporal information can also be of value in classifying SAR images. For example (Cihlar *et al.* 1992) a variety of land-cover types (including agricultural, urban, forestry and golf courses) display distinctive annual sinusoidal variations in backscatter and consistent year-to-year stability in backscatter profiles. To date there has been insufficient satellite SAR to exploit this phenomenon but the current and planned missions together with the ability to program SAR images 'to order' could see great advances in multitemporal techniques for classifying radar images. Figure 5.17 shows experimental results using C-band aerial SAR which demonstrate how it is possible to distinguish land-cover classes on the basis of annual backscatter range combined with annual mean backscatter.

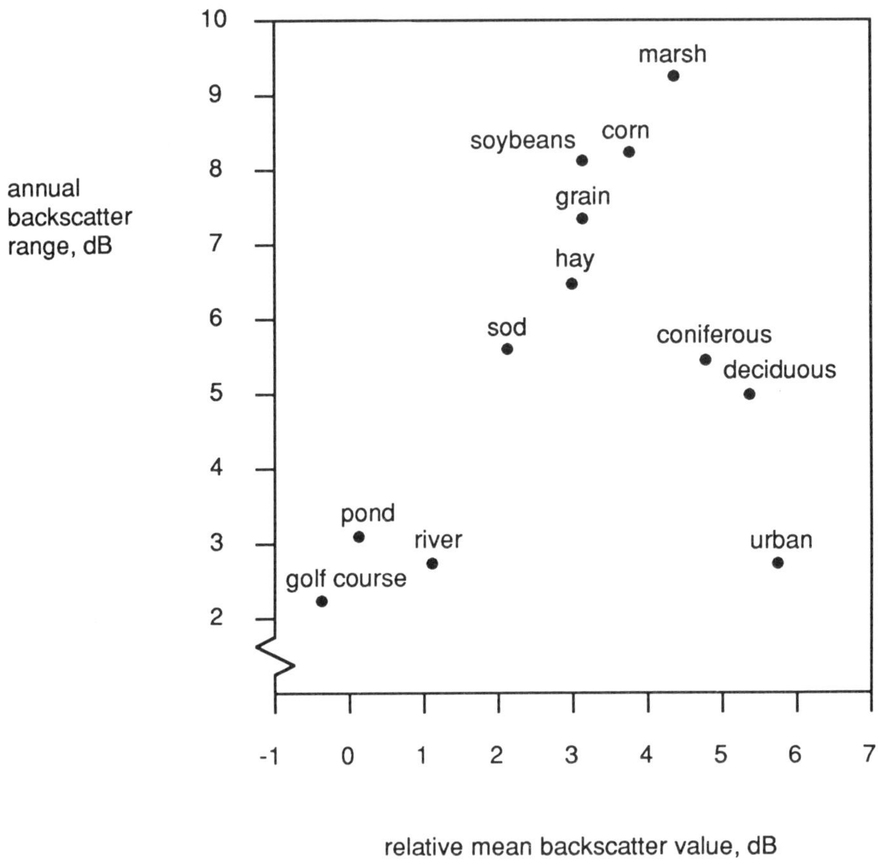

Fig. 5.17. Distinguishing land cover classes on the basis of annual backscatter range combined with annual mean backscatter—experimental results derived from aircraft C-band SAR (based on Cihlar *et al.* 1992)

5.4 CHANGE DETECTION

Change detection is concerned with identifying and quantifying areas of change, for example deforestation or urban growth. Other potential applications include study of agricultural changes, monitoring of coastal processes and hazard and disaster assessment. To compare two or more images of the same area it is essential that they can be related to a single coordinate system. For interactive or cerebral processing it is often sufficient that the images are registered together using one image (typically the earliest one) as the base image (see section 1.1.6 for details). For example figure 5.18 (in the plate section between pp. 84 and 85) shows three ERS-1 SAR images of Morecambe Bay in England, these have been assigned to an RGB composite described in table 5.9. The pink areas represent the tidal flats and sandbanks exposed at low tide and show a great variety of change compared with the more stable inland areas which appear mainly in shades of grey with some more colourful patches due to the ripening and harvesting of certain agricultural crops.

Table 5.9. ERS-1 SAR composite image (courtesy Radarsat International, ESA and Liverpool Coastguard)

Band	Date	Tidal state
Blue	1 August 1991	40 minutes after low tide
Green	7 August 1991	20 minutes after high tide
Red	13 August 1991	130 minutes after low tide

For more accurate multitemporal analysis, it is desirable that the images are all geocoded to the same map-projection. Ideally the geocoding process should result in an accuracy of better than one pixel in registration. If this is not achieved then small areas or edges of larger areas may be detected as spurious change due to the inaccuracy of the registration. Even when using accurate models of satellite and terrain geometry it is not always possible to achieve the desired accuracy in a first-pass geocoding process and it may be necessary to fine tune the registration using further tie points or correlation based methods. It is also important to produce a multitemporal series whose radiometric values are directly comparable—change must be attributable to genuine surface changes and not to systematic effects such as varying illumination conditions, atmospheric attenuation or changes to sensor gain or bias. A simple method of doing this is by histogram matching; this will work fairly well providing that the areas of change are not large enough to bias the histograms significantly. However, as with all modelled interpretation of satellite data, the most accurate results will be obtainable if the data is accurately calibrated and converted to geophysical units.

After the images have been precision geocoded and calibrated it can be hypothesised that the changes between the two images can be attributed to genuine terrestrial changes. An initial estimate of the degree of change between the two images can be performed using an IIPF. One technique (for example) is to display one image on a red gun and the second on a green gun. Areas of no change will appear as grey-tones with areas of great change being visible in one of the primary colours. Other techniques include switching rapidly from one image to the other whereby areas of great change appear the most flickery.

5.4.1 Image differencing

More quantitative results can be achieved by using algorithmic methods—the simplest being to subtract the two images. This results in a 'change image' whose values are defined as follows:

$$C(i,j) = I_2(i,j) - I_1(i,j) \tag{5.5}$$

where

$C(i,j) =$ value of pixel (i,j) in change image,

$I_1(i,j) =$ value of pixel (i,j) in oldest image,

$I_2(i,j) =$ value of pixel (i,j) in youngest image.

A typical histogram of such a change image can be seen in figure 5.19; this can be interpreted as follows: The mode present around zero is contributed by the areas with no significant change; the tails of the histogram correspond to the areas of greatest change.

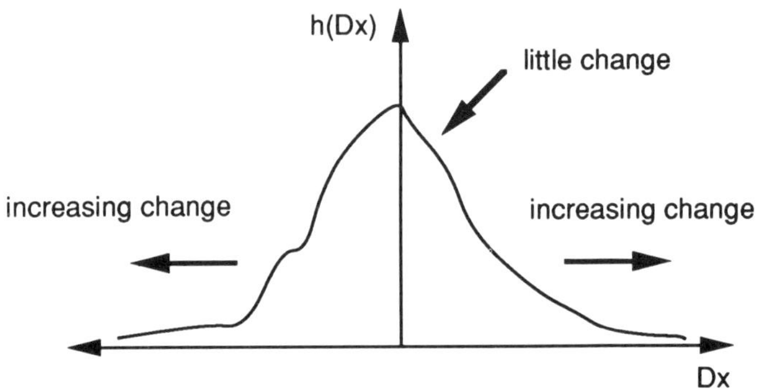

Fig 5.19. Schematic histogram of a change image derived by subtracting two registered multitemporal images.

The change image itself can be interpreted using two filtering techniques:

(1) thresholding of the image histogram—this removes the pixels near the centre of the histogram which are not associated with change;
(2) cleaning of small areas—to remove isolated pixels and edge-pixels where the detected change may be caused spuriously, for example by inaccurate registration, or to remove pixels that suffer from mixture effects.

The result is a number of areas whose size and spatial relationships can be measured. This may not in itself be enough to identify change but it may at least provide suitable 'triggers' for future examination.

5.4.2 Other change detection algorithms

A number of other change detection techniques have been proposed and a survey can be found in the review article by Ashbindu Singh (1989). The first class of algorithms are essentially variants on the image differencing technique and include image ratioing and regression analysis. The advantage of image ratioing is that it detects large relative changes in low pixel-values that would be thresholded-out by image differencing; regression analysis uses statistical analysis to find the best-fit line between pixel values in the two images, effectively eliminating differences in the mean value and variance from the two dates.

A second class of algorithms is concerned with comparing transforms of the image data; suitable choices include vegetation index transforms and the principal components transform (PCT). Vegetation index methods can make use of any of the algorithms described in section 5.2.2, such as the perpendicular vegetation index (PVI) or the tasselled-cap transform. The comparison between the two transformed images from the different dates can then be performed by using simple image differencing. A similar approach is used for the PCT images; alternatively a principal components transform (see section 5.2.1) can be performed on the entire data set amalgamated from the two images. The expected result (Byrne *et al.* 1980) is that major changes associated with scene illumination and atmospheric effects should appear in the major components with the 'interesting' changes being present in the minor components. It is however relatively difficult to analyse such a data set.

The third major approach is to perform a multispectral classification on each image individually and to compare the results, producing a class transition for each pixel in the image. This approach is best used within the context of a geographic information system (see chapter 6 for further details).

5.5 APPLICATIONS

5.5.1 Renewable resources

With an increasing world population, effective use of renewable resources is becoming ever more vital. Satellite data is proving increasingly useful for the monitoring of these resources, particularly in agriculture and natural vegetation and forestry. The classification of agriculture crops was one of the earliest applications of Landsat-MSS in the early 1970s (Wiegand *et al.* 1973; Hall *et al.* 1974), the most prominent example of which was the Large Area Crop Inventory Experiment (LACIE) carried out between the US National Oceanographic and Atmospheric Administration (NOAA), the US National Aeronautics and Space Administration (NASA) and the US Department of Agriculture. Multispectral classification remains the primary method of deriving local vegetation information from space (the derivation of global-scale vegetation information is described in section 5.5.3) and has been extensively reported in the remote sensing literature. Rather than attempt a comprehensive review this section concentrates on two specific examples:

(1) the successes that have been achieved using the series of Indian Remote Sensing satellites (IRS) which provides an important example of a developing country using remote sensing as a practical solution to the management of renewable resources (Kasturirangan *et al.* 1991);

(2) the application of remote sensing to forestry management.

Indian remote sensing satellites

India suffers from many of the problems associated with underdeveloped countries; it is densely populated and its 700-plus million inhabitants are greatly dependent on agriculture for a living (Rao 1991). The increased agricultural productivity has resulted in large-scale irrigation of over 30 per cent of the sub-continent, compared with a typical figure of less than 10 per cent in many 'developed' nations. However large-scale irrigation, coupled with extensive use of chemical fertilisers, poor drainage and inappropriate agricultural practices has led to an undesirable increase in soil salinity, reducing previously fertile land to an impoverished and unproductive state. India is still inadequately mapped and therefore there is a great need for the collection of information on the country's natural resources and their subsequent effective management. A key part of this process has been the continued role played by the development of India's own national remote sensing program, culminating in the launch of Indian Remote-sensing Satellite-1 (IRS-1) in 1988 and forming the central component in an integrated Indian national natural resources management system encompassing not only agriculture but general land use, geology, hydrology, flood monitoring, snow and ice monitoring and marine studies. A summary of the IRS-1 technical specification can be seen in table 5.10.

Table 5.10. IRS-1 technical specification

Country of origin	India
Launch	17 March 1988
Orbit	sun-synchronous (near polar)
Orbit period	103 minutes
Inclination	99°
Altitude	904 km
Coverage	81° N to 81° S
Repeat cycle	22 days
Payload	Linear Imaging Self Scanner (LISS)-I Linear Imaging Self Scanner (LISS)-II
Swath width	148 km (LISS-I) 145 km (LISS-II)
Pixel size	72 m (LISS-I) 36 m (LISS-II)
Band 1	0.45–0.52 μm
Band 2	0.52–0.59 μm
Band 3	0.62–0.68 μm
Band 4	0.77–0.86 μm

The objective of IRS-1 is to provide an Indian 'end-to-end' system with India designing, building and operating the satellite, receiving and processing the data and applying the results to the pressing needs of national development. The first step taken towards this objective was the 'Bhaskara' experimental program carried out during 1976–1982. The Bhaskara platform featured two types of remote sensing instruments—a slow-scan vidicon camera operating in the visible and near-infra-red with a resolution of 1 km and various microwave radiometers. Data was used to provide a land cover and geolithological mapping of India at 1:2 000 000; relatively coarse but still much improved on previously available information. Based on the experience of the Bhaskara experimental program it was decided to embark on a higher-resolution pre-operational system. The objectives of the IRS-1 program were to be threefold:

— to design and develop an indigenous remote sensing satellite;
— to develop operational support in the ground segment;
— to use data in a complementary fashion with that from more conventional sources.

Since its launch IRS-1 has provided data for many successful applications in renewable resources as well as other applications such as geology, wasteland mapping and flood damage assessment. One of its most important applications is the search for ground-water. India, in common with much of the developing world, needs adequate sources of water for agricultural development and industry as well as drinking and domestic use. Originally, Landsat data was used (Sahai *et al.* 1985) to help detect the hydrological and geological features that are indicative of ground-water occurrence. IRS-1 data was subsequently used (Sahai *et al.* 1991) to prepare hydro-geomorphological maps, from which potential ground-water sites around 'problem' villages can be located. Work carried out in various areas resulted in a success rate of over 90% in most cases (compared with typical results of 60% in cases where satellite data has not been used).

Forestry management

Forestry is also of great importance to the economic welfare of many nations and satellite images have provided valuable input to the regular and detailed forest inventories that are necessary to manage forests effectively. On a global scale, satellite data also has an important role in monitoring the effect that deforestation can have on the delicately balanced global model (this topic is described in more detail in section 5.5.3). One application of satellite imagery to forestry (Keil *et al.* 1992) was the use of Landsat-TM for forest classification and dead wood mapping in the largest woodland in central Europe, a conglomeration of the national park of the Bavarian forest in Germany and the national park of Sumava (the Czech Republic/Bohemia). Remote sensing can be used to monitor this environmentally sensitive region and to map 'areas of retreat' for animals and plants along the border region, which had been little-mapped prior to the raising of the iron curtain.

Another forestry application was the use of SPOT images to assess storm damage in forests (Leysen et al. 1992). Storm damage is particularly important in a European context, for example in Flanders (Belgium) the equivalent of the annual harvest (700 000 m^3, or 1.7% of the forested area in Flanders) was destroyed in just three days of storm damage in January and February 1990. Another regional forestry project (Brown and Fox 1992) using satellite data was a project to map timberland using Landsat-TM in Klamath Province, California. This work was a consequence of a California state assembly bill to form a 'timberlands task force' to improve protection of wildlife resources. The aim is to develop rules to identify wildlife habitat relationships (WHR) cover types over a six million acre study area. The classification scheme was based on a number of attributes such as vegetation classes, canopy closure and tree size classes; for example the canopy closure classes are shown in table 5.11. The WHR vegetation classes can be seen in table 5.12.

Table 5.11. WHR tree canopy closure classes (based on Brown and Fox 1992)

Class	Canopy closure
0	non-vegetation types
1	10–24% (sparse)
2	25–39% (open)
3	40–59% (moderate)
4	≧ 60% (dense)

Table 5.12. WHR vegetation classes (based on Brown and Fox 1992)

subalpine conifer	eastside pine
red fir	Pinyon juniper
white fir	juniper
Douglas fir	mixed conifer
redwood	hardwood
closed-cone pine/cypress	shrub
montane hardwood/conifer	barren
lodgepole pine	herbaceous
Jeffrey pine	water
Ponderosa pine	

5.5.2 Non-renewable resources

It is becoming increasingly important to find new methods to detect scarce geological and mineral resources. Optical/IR satellite images can be used to help detect and classify non-renewable resources directly from spectral signature (if the rock or soil is exposed). Alternatively SAR may be used due to its ability to penetrate vegetation and surface layers and its sensitivity to surface roughness and moisture content.

Satellite images may also provide secondary clues such as soil moisture and vegetation changes. This latter approach, known as geo-botany, is used where it is not possible to sense mineral deposits directly because of vegetation and makes use of the analysis of plant species, their distribution and health as a proxy indicator of the underlying mineral. Geo-botanical anomalies that may indicate location of mineral deposits (Mather 1987) include:

— reduced ground cover or stunted growth;
— unexpected species distribution;
— yellowing or other pigmentation changes in leaves;
— changes in the phenological cycle (such as late leafing or early senescence).

Using direct methods, satellite images have proved an important contribution to extraction of geological information. As well as techniques of multispectral classification a number of other algorithms have proved to be more specifically suited to geological applications. These include:

— band ratios;
— the IHS transform;
— the principal components transform;
— lineament mapping.

Band ratios

The motivation for using band ratios for geological purposes is similar to that for computing vegetation indexes (see section 5.2.2) in that by comparing the relative value of two bands the need for accurate absolute calibration, and accounting for the topographic effect, become less paramount. Various ratios can be employed depending on the particular application, for example (Trefois and Volon 1992) a colour composite made up of Landsat-TM band ratios 5/7, 5/1 and (5+3)/4 was used to emphasise hydroxyl and carbonate band absorptions in a lithologic mapping of Niger.

The IHS transform

The use of the IHS transform has already been described for visual enhancement of images. An example of the use of the IHS transform for geological mapping is a project to map hydrothermally altered rocks with Landsat data (US Geological Survey 1988). Areas of such rocks can indicate the presence of hydrothermal mineral deposits, for example limonite anomalies can be associated with high levels of gold, silver, copper and other minerals. The Landsat sensor is particularly suitable for limonite mapping as its spectral range encompasses two characteristic limonite absorption features. The USGS project carried out in Colorado and Saudi Arabia successfully used satellite images as part of a multi-disciplinary metallic mineral assessment. The spectral difference in the images used were enhanced by image ratioing and by colour coding using an IHS transform. The results were then subjected to further interpretation by geologists to distinguish hydrothermally altered limonite from that which may have been subject to other surface processes. The IHS transform was also used in a study of mineral exploration in the English Lake District (Forrest and Harding 1988). In this case Landsat-TM data was used as a single 'band' in a multisource data set that also comprised digital maps of regional geochemistry and geophysics.

Principal components transform

The principal components transform (PCT) has also proved of use for enhancing the value of images for geological analysis; for example the PCT was one of the techniques for drainage network analysis of Mt. Olympus in Greece, using Landsat-TM images (Astaras and Soulakellis 1992). In this case the first three principal components of a Landsat-TM image from July 1987 were found to contain 98.6% of the information (variance) of the set of six TM bands (excluding thermal band 6). Drainage lines proved to be more visible in the first principal component image (PC1) than in any of the original data set.

Lineament mapping

The mapping of lineaments is of importance geologically and a number of filters may be used to extract edges (points with rapid spatial change in grey-level). Whilst such methods are generally satisfactory for locating candidate edges, the knowledge-based approach described in section 6.3.1, may be necessary to associate individual edge pixels with the linear features of interest. The Hough transform (see section 5.2.3) is also useful for detection of geological lineaments. It can however be used for any shape that can be parameterised, for example (Cross 1988) it can be used to detect circular geological features such as granite intrusions, salt domes, impact craters and volcanic features. In this case the standard equation for a circle is used:

$$(x-a)^2+(y-b)^2=r^2 \tag{5.6}$$

where

$a =$ x-coordinate of circle centre,

$b =$ y-coordinate of circle centre,

$r =$ circle radius.

Because there are three parameters (a, b, and r) a three-dimensional accumulator array is used.

5.5.3 Global monitoring

For many parts of the developing world such as Africa and Asia it is vital to be able to maintain records of land cover, especially vegetation, over wide regions and with a rapid update rate in order to follow the dynamics of vegetation during the growing season. Before the advent of satellites the limited resources available for ground and aerial surveys in such countries often proved inadequate to contribute to the planned and effective use of the available natural resources necessary for sustained social and economic growth. A second imperative for the reliable classification of land cover at continental and even global scales is the impact of land-cover dynamics on the global energy balance, the hydrological cycle and the production of carbon dioxide.

Data sources previously available have proved inaccurate, inconsistent and difficult to keep up to date—for example global estimates of tropical rain forest areas compiled in 1979 differed from each other by up to 50%. Since the availability of Landsat images in 1972 it has at least been possible to conceive of a practical earth resources monitoring program using satellite data, but the transfer of remote sensing technology from the research sector to operational use has proved difficult in practice. One obvious factor is the sheer volume of data involved—for example the continental-scale coverage of Africa would require of the order of 1000 Landsat scenes. The amount of processing required to obtain suitable images and process them on a one-off basis would be phenomenal—to do so on a repetitive monitoring basis would be almost impossible with current technology.

A realistic alternative is the use of AVHRR-derived vegetation indexes (see section 5.2.2). Amongst the topics in vegetation mapping that may be addressed by these methods are:

— detection and monitoring;
— classification;
— evaluation of vegetation vigour;
— estimates of biomass.

In the longer term such information could prove vital as input to models which relate the vegetation cover of the earth to geochemical, hydrological and climatic cycles (see section 6.4.1). Probably the most used vegetation index is the NDVI. Typical NDVI values determined from examining the global area coverage (GAC) of land-cover categories such as oceans, lakes, deserts, grassland, and forests can be seen in table 5.13.

Table 5.13. Typical NDVI values (based on Holgen 1986). These values were derived from NOAA-7 data.

Land cover category	Red	Infra-red	NDVI
Green-leaf vegetation			
dense	0.050	0.150	0.500
medium	0.080	0.110	0.140
light	0.100	0.120	0.090
Bare soil	0.269	0.283	0.025
Clouds	0.227	0.228	0.002
Snow/ice	0.375	0.342	−0.046
Water	0.022	0.013	−0.257

An early example of the utility of AVHRR for large-scale vegetation monitoring (Malingreau 1986) was the use of weekly data to analyse a wide variety of vegetation formations ranging from the deserts of Central Asia to the tropical forests of Borneo. Development curves of the NDVI index over the period 1982–1985 demonstrated the large amount of useful information that can be derived. AVHRR data also formed the basis of NDVI calculations for the whole of Africa sampled at 8 km resolution (Townshend and Justice 1986). In this case temporal compositing of the images was used to reduce the effects of cloud cover and atmosphere. Areas of different land-cover classes are distinguishable by multitemporal analysis of the changing NDVI over a year. For example, deserts have a very low NDVI all year round but tropical rainforests have a much higher NDVI as well as marked seasonal peaks (see figure 5.20).

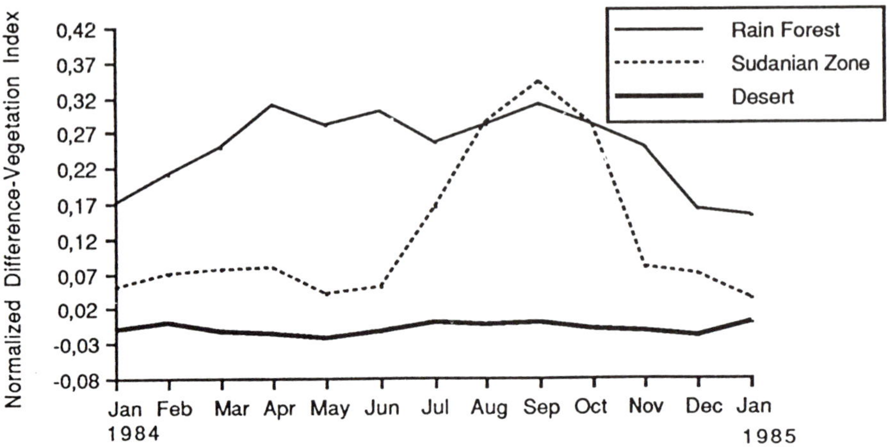

Fig. 5.20. NDVI seasonal development curves for three African land-cover types in 1984 (from Belward 1991).

5.5.4 Other applications

There are many wide-ranging applications of satellite images including oceanography, atmospheric monitoring, urban studies and the collection of military intelligence. It is not the intention of this book to provide a comprehensive review of thematic mapping from space; instead two sample application areas are described:

— hazard monitoring;
— bathymetry and coastal studies.

Hazard monitoring

The monitoring of hazards is of increasing importance to the welfare of mankind on this planet, for example (Vinogradov and Frolov 1992) it is estimated that ecological disasters currently effect more than 16–18% of the land area of Russia. The definition of a hazard is

> '... a condition capable of exerting adverse effects on a human life, property or activity' (Barrett and Beaumont 1992)

whereas a disaster is a serious life-threatening situation which has arisen when a hazard has exceeded some critical limit. Natural disasters include hurricanes and floods, whilst well-known man-made environmental disasters include the pollution caused by the Gulf War, the EXXON-Valdez tanker spill (Jensen *et al.* 1990) and the accident at the Chernobyl nuclear reactor. Typical categories of disaster are shown in table 5.14.

Remote sensing is typically of use for three types of hazard monitoring:

Table 5.14. Categories of disaster (based on Montaperto and Rossi 1992)

Category	Description
Geological	earthquake volcanic eruption desertification soil erosion
Vegetation	forest fire deforestation locust infestation
Weather	flood snow drought hurricane
Air pollution	chemical radioactive
Water pollution	sea pollution pollution of inland water

(1) to monitor the 'steady state' of a potential hazard and to assist in disaster prediction (for example by detection of anomalous behaviour);
(2) when a disaster has occurred to provide an assessment of the evolving situation;
(3) after a disaster to determine the effect of remedial action and to reassesses the status of the hazard.

One specific example of the utility of satellite images for hazard monitoring was the detection of snow-melt on the top of the Nevado Sabencaya mountain in Columbia (Abiodun 1992) which provided warning of the imminent eruption of the volcano which took place on 14 July 1990. Other examples of applications of remote sensing for hazard monitoring include:

(1) observing volcanic ash clouds using AVHRR data (Prata 1989);
(2) production of an inundation risk map for areas of the Buenos Aires province (Argentina) susceptible to flooding (Mehl and Hiller 1992);
(3) analysis of the effects of human activities on the desertification and degradation of the Sahel grasslands in Mali (Csaplovics 1992).

Because it is important to be able to observe the evolution of hazard-related phenomena through time the ability to acquire and interpret multitemporal image sets of a hazard region is of paramount importance. For example a three-year time series of Landsat-TM and SPOT-XS images was used to analyse areas devastated by fire on the island of

Sardinia (Marini and Mellis 1992). Multitemporal Landsat images were also used in a study of flood damage to paddy fields in the Kanto district of Japan (Yamagata and Akiyama 1988). The image analysis flowlines used are illustrated in figure 5.21.

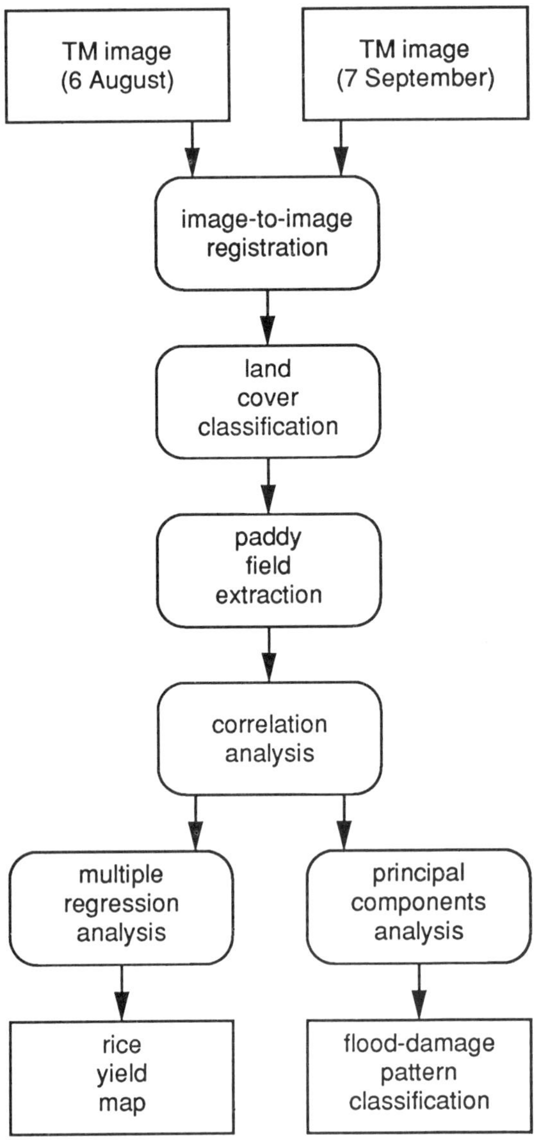

Fig. 5.21.Operational flowlines for a study of flood damage to paddy fields in the Kanto district of Japan (adapted from Yamagata and Akiyama 1988).

Bathymetry and coastal studies

It is of great importance to the world's naval and shipping authorities to have access to accurate and frequently revised naval charts. The important features of such charts can be divided into two main classes—surface topographic features such as islands and coastlines and bathymetric features below the sea surface. For much of the world's sea areas such charts are unavailable, unrevised or were wrongly drawn up in the first place, for example as the result of an inaccurate depth sounding. Thus many of the hazards to world shipping are not mapped in sufficient detail for modern needs. In ancient times man's knowledge of the sea-bed was derived simply by use of rods and weighted lines; both methods severely limited in locality and to relatively shallow areas. In the twentieth century echo-sounding equipment became widespread until it was superseded by the technically superior sonar systems in common use today. Even with such methods

> '... it is still true that the world's largest vessels pass through waters last surveyed a hundred years ago or more' (Benny and Dawson 1983).

As with many of the land-based problems described previously, remote sensing can play a major contribution to topographic mapping, and bathymetry from satellite (although limited to shallow water applications) can be used to provide valuable input to nautical charts. Optical imagery has great advantages in three areas of nautical chart preparation:

(1) augmenting and updating data which already exists for an area ('putting new life into old surveys');
(2) as a cost-effective aid to planning hydrographic surveys;
(3) monitoring mobile areas of sea-bed which are not currently charted due to their dynamic nature.

Shallow-water bathymetry can be carried out by considering the various routes that light can reach the satellite sensor (see figure 5.22):

— reflection from the surface;
— reflection from the bottom;
— reflection or scattering from particles or objects within the water.

Mirror-like (specular) reflection is also possible but only if the sea surface is almost unbroken. If the water is clear and of shallow depth the light can be returned and detected by the satellite. Light return depends on the depth of the water, the attenuation coefficient for that wavelength and the reflection coefficient of the bottom. Attenuation is characteristically low in visible bands and high in the infra-red and in general can be expressed by:

$$I = I_0 \exp(-dK) \qquad (5.7)$$

where

I = amount of light remaining after attenuation by water,

I_0 = strength of incident light,

d = depth in metres,

K = attenuation coefficient.

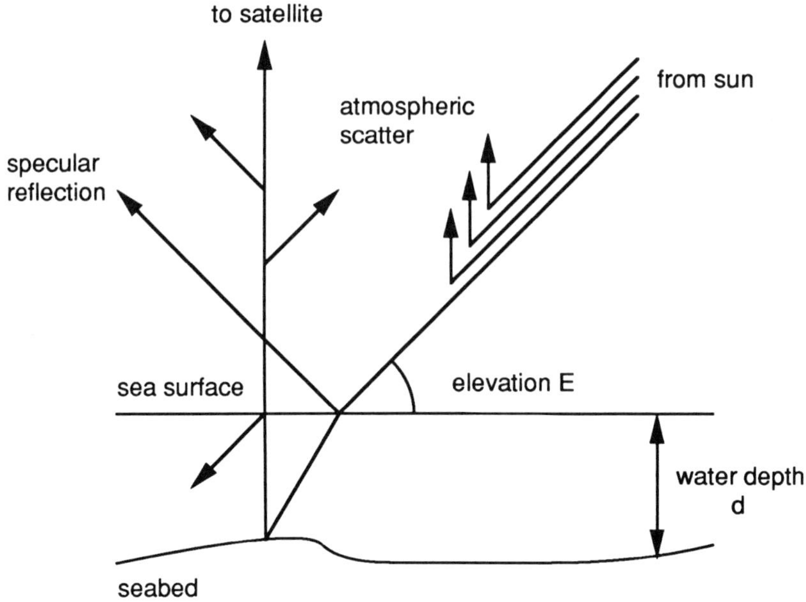

Fig. 5.22. Basic imaging geometry used in bathymetry (adapted from Benny and Dawson 1983).

This is the method used (Benny and Dawson 1983) to map a region between Ras Wadi Tiryam and Ras al Fasma in the northern Red Sea, using Landsat-MSS images. The processing approach adopted was to calculate the depth maps from raw un-geocoded data and to transform the final depth contours to a map projection. There are, however, limitations to the utility of satellite images for bathymetric mapping. These are primarily:

(1) depth limitations (typically 20 m in clear water);
(2) poor spectral and spatial discrimination: for navigation purposes it is vital to be able to locate pinnacle features, no matter how narrow they are;
(3) inaccuracies in depth measurements.

Another application for which satellite data can successfully be used is the study of coral reefs (Kuchler *et al.* 1986), for example reef geography, form, surface cover and vegetation. In particular remote sensing techniques have been applied to the management of the Great Barrier Reef (GBR) which lies on the north-east continental shelf of Australia. The reef, designated a world heritage site, is one of the most diverse ecosystems on earth comprising a series of about 2500 reefs with a total length of about 2000 km. The area of the Great Barrier Reef Marine Park is approximately 345 000 km^2, thus providing a very large object for local survey. Interest in using satellite data as a possible alternative to conventional mapping (Harrison *et al.* 1989) led to the initial development of the 'Barrier Reef Image Analysis' package (BRIAN) in 1983. This was used to produce geocoded image maps for the whole of the barrier reef at a scale of 1:250 000 using standard multispectral classification techniques, as well as 'depth of penetration' algorithms using Landsat-TM data.

5.6 REFERENCES

Abiodun A A (1992) Natural disasters and their mitigation using space technology. *Proceedings of the Central Symposium of the 'International Space Year' Conference*, Munich, Germany, 30 March–4 April 1992, ESA ISY-1, 573–580.

Astaras T A and Soulakellis N A (1992) Contribution of digital image analysis techniques on Landsat-5 TM imageries for drainage network delineation. A case study from the Olympus mountain, W. Macedonia, Greece, Remote Sensing From Research to Operation. *Proceedings of the 18th Annual Conference of the Remote Sensing Society*, University of Dundee, Scotland, 15–17 September 1992, 163–172.

Barrett E C and Beaumont M J (1992) Alternative strategies for the use of satellite data analysis systems for hazard monitoring and disaster mitigation. *Proceedings of the Central Symposium of the 'International Space Year' Conference*, Munich, Germany, 30 March–4 April 1992, ESA ISY-1, 581–586.

Belward A S and Valenzuela C R (eds.) (1991) *Remote Sensing and Geographical Information Systems for Resource Management in Developing Countries*. Kluwer Academic Publishers.

Benny A H (1980) Coastal definition using Landsat data. *International Journal of Remote Sensing*, **1** (3) 255–260.

Benny A H and Dawson G J (1983) Satellite imagery as an aid to bathymetric charting in the Red Sea. *The Cartographic Journal*, **20** (1) 5–16.

Bird A C (1991) Principles of remote sensing: interaction of electromagnetic radiation with the atmosphere and the earth. In Belward and Valenzuela (eds.) (1991), 17–30.

Brown G K and Fox L (1992) Digital classification of Landsat Thematic Mapper imagery for recognition of wildlife habitat characteristics. *Proceedings ASPRS/ACSM Convention, American Society for Photogrammetry and Remote Sensing*, Bethesda, Maryland, 1992, 251–260.

Byrne G F, Crapper P F and Mayo K K (1980) Monitoring land cover change by principal component analysis of multitemporal Landsat data. *Remote Sensing of Environment*, **10**, 175–184.

Cihlar J, Pultz T J and Gray A L (1992) Change detection with Synthetic Aperture Radar. *International Journal of Remote Sensing*, **13** (3) 401–414.

Chavez P S (1988) An improved dark-object subtraction technique for atmospheric scattering correction of multispectral data. *Remote Sensing of Environment*, **24**, 459–479.

Churchill P N and Sieber A J (1991) The current status of ERS-1 and the role of radar remote sensing for the management of natural resources in developing countries. In Belward and Valenzuela (eds.) (1991), 111–143.

Cornaert M-H and Maes J (1992) Land cover an essential component of the CORINE information system on the environment. GIS Implications. *Proceedings of the Central Symposium of the 'International Space Year' Conference*, Munich, Germany, 30 March—4 April 1992, ESA ISY-1, 473–481.

Crist E P and Cicone R C (1984) A physically-based transformation of Thematic Mapper data—the TM tasseled cap. *IEEE Transactions on Geoscience and Remote Sensing*, **GE-22** (3) 256–263.

Cross A M (1988) Detection of circular geological features using the Hough transform.

International Journal of Remote Sensing, **9** (9) 1519–1528.

Csaplovics E (1992) Analysis of human impact on degradation and desertification of the West African Sahel by means of multi-level remote sensing techniques. *Proceedings of Satellite Symposia 1 and 2 : Navigation and Mobile Communications, and Image Processing, GIS and Space-assisted Mapping, from the 'International Space Year' Conference*, Munich, Germany, 30 March–4 April 1992, ESA ISY-2, 413–416.

Dengere J (ed.) (1988) *Thematic mapping from satellite imagery. An International Report.* Elsevier Applied Science Publications, Barking.

Drake N and White K (1991) Linear mixture modelling of Landsat Thematic Mapper data for mapping the distribution and abundance of gypsum in the Tunisian Southern Atlas. *Proceedings of Spatial Data 2000*, Christ Church, Oxford, 17–20 September 1991, 168–177.

Foody G M (1986) An assessment of the topographic effects on SAR image tone. *Canadian Journal of Remote Sensing*, **12** (2) 124–131.

Foody G M (1987) A method for thematic classification with Synthetic Aperture Radar data. *Geocarto International*, **2**, 31–36.

Foody G M (1992) Classification accuracy assessment: some alternatives to the Kappa coefficient for nominal and ordinal level classifications. *Proceedings of the 18th Annual Conference of the Remote Sensing Society*, University of Dundee, Scotland, 15–17 September 1992, 529–538.

Foody G M and Cox D P (1991) Estimation of sub-pixel land cover composition from spectral mixture models. *Proceedings of Spatial Data 2000*, Christ Church, Oxford, 17–20 September 1991, 186–195.

Forrest M D and Harding A E (1988) Integration of Landsat TM, stream sediment geochemistry and regional geophysics for mineral exploration in the English Lake District. *Proceedings of the Sixth Thematic Conference on Remote Sensing for Exploration Geology : Applications, Technology, Economics*, 16–19 May 1988, Houston, Texas, 469–474.

van Genderen J L and Uiterwijk U (1987) A practical procedure for classifying digital imagery. *Proceedings of the Annual Conference of the Remote Sensing Society*, Nottingham, September 1987, 287–296.

Hall F G, Bauer M E and Malila W A (1974) First results from the Crop Identification Technology Assessment for Remote Sensing (CITARS). *Proceedings of the Ninth International Symposium on Remote Sensing of Environment*, Ann Arbor, Michigan, 1171–1191.

Harrison B A, Jupp D L B, Hutton P G and Mayo K K (1989) Accessing remote sensing technology, the microBRIAN example. *International Journal of Remote Sensing*, **10** (2) 301–309.

Hayes L (1985) The current use of TIROS-N series of meteorological satellites for land-cover studies. *International Journal of Remote Sensing*, **6** (1) 35-45.

Holben B N (1986) Characteristics of maximum value composite images from temporal AVHRR data. *International Journal of Remote Sensing*, **7** (11) 1417.

IEEE (1976) Standard letter designation for radar frequency bands. IEEE standard 521-1976.

Jensen J R, Ramsey E W, Holmes J M, Michel J E, Savitsky B and Davis B A (1990) Environmental Sensitivity Index (ESI) mapping for oil spills using remote sensing and Geographic Information System technology. *International Journal of Geographical Information Systems*, **4** (2) 181–201.

Jupp D L B and Mayo K K (1982) Use of residual images in Landsat image analysis. *Photogrammetric Engineering and Remote Sensing*, **48**, 595–604.

Kasturirangan K, Joseph G, Kalyanraman S, Thyagarajan K, Chandrasekhar M G, Raju D V, Raghunathan S, Gopalan A K S, Venkatachari K V, Shivakumar S K (1991) IRS mission. *Current Science (India)*, **61** (3 and 4) 136–151.

Kauth R J and Thomas G S (1976) The tasseled cap—a graphic description of the spectral temporal development of agricultural crops as seen by Landsat. *Proceedings of the Second Annual Symposium on Machine Processing of Remotely Sensed Data*, Purdue University, Indiana, 1976, 4B41–4B49.

Keil M, Coenradie B, Rall H and Sima M (1992) The national parks of Bavarian Forest and Sumava—a Landsat TM perspective on the structure and dynamics of the largest woodland of central Europe. *Proceedings of the Central Symposium of the 'International Space Year' Conference*, Munich, Germany, 30 March–4 April 1992, ESA ISY-1, July 1992, 751–755.

Kuchler D A, Jupp D L B, Claasen D B R and Bour W (1986) Coral reef remote sensing applications. *Geocarto International*, **4**, 3–15.

Lee J-S (1981) Refined filtering of image noise using local statistics. *Computer Graphics and Image Processing*, **15**, 380–389.

Leysen M M, de Roover B P, Goossens R E and de Wulf R R (1992) Assessment of storm damage in forests using SPOT imagery. *Proceedings of the Central Symposium of the 'International Space Year' Conference*, Munich, Germany, 30 March–4 April 1992, ESA ISY-1, 773–777.

Lintz J and Simonett D S (eds.) (1976) *Remote sensing of environment*. Addison-Wesley Publishing Company, Reading, Massachusetts, 1976.

Malingreau J-P (1986) Global vegetation dynamics: satellite observations over Asia. *International Journal of Remote Sensing*, **7** (9) 1121–1146.

Marini A and Mellis M T (1992) TM-SPOT multitemporal analysis on areas devastated by fire. *Proceedings of Satellite Symposia 1 and 2: Navigation and Mobile Communications, and Image Processing, GIS and Space-assisted Mapping, from the 'International Space Year' Conference*, held in Munich, Germany, 30 March–4 April 1992, ESA ISY-2, 423–425.

Mather P M (1987) *Computer processing of remotely-sensed images: an introduction*. John Wiley and Sons, Chichester.

Mehl H and Hiller K (1992) Application of multitemporal Landsat-MSS and TM data for the development of an inundation-risk-map. *Proceedings of the Central Symposium of the 'International Space Year' Conference*, Munich, Germany, 30 March–4 April 1992, ESA ISY-1, 591–594.

Montaperto A T and Rossi F (1992) Use of remote sensing spacecrafts in real time support to disaster management: interfaces with ground segment facilities and services. *Proceedings of the Central Symposium of the 'International Space Year' Conference*, Munich, Germany, 30 March–4 April 1992, ESA ISY-1, 599–604.

Prata A J (1989) Observations of volcanic ash clouds in the 10–12 μm window using AVHRR/2 data. *International Journal of Remote Sensing*, **10** (4 and 5) 751–761.

Pratt W K (1985) *Digital image processing*. John Wiley, New York.

Quegan S, Churchill P N, Wright A, Lamont J, Rye A J and Trevett J W (1985) SAR for agriculture and forestry. *Proceedings of a Workshop on Thematic Applications of SAR Data, ESRIN*, Frascati, 9–11 September 1985, ESA SP-257, 7–14.

Rao U R (1991) Space and agriculture management. In 'Space and Agriculture Management', Special Current Event Session, International Astronautical Federation, 42nd IAF Congress, Montreal, Canada, 1991, 1–10.

Richardson A J and Wiegand C L (1977) Distinguishing vegetation from soil-background information. *Photogrammetric Engineering and Remote Sensing*, **43**, 1541–1542.

Sabins F F (1987) *Remote sensing principles and interpretation*. Second edition. W H Freeman and Company, New York.

Sahai B, Bhattacharya A and Hegde V S (1991) IRS-1A applications for groundwater targetting. *Current Science (India)*, **61** (3 and 4) 172–179.

Sahai B, Sood R K and Sharma S C (1985) Ground water exploration in the Saurashtra peninsula. *International Journal of Remote Sensing*, **6** (3 and 4) 433–441.

Settle J J (1990) Contextual models and the use of Dirichlet priors for mixing problems. *Proceedings of the Fifth Australian Remote Sensing Conference*, Perth, Western Australia, 8–12 October 1990, 314–323.

Sheffield C (1985) Selecting band combinations from multispectral data. *Photogrammetric Engineering and Remote Sensing*, **51** (6) 681–687.

Simonett D S, Strahler A H, Sun G and Wang Y (1987) Radar forest modeling: potentials, problems, approaches, models. *Proceedings of the Annual Conference of the Remote Sensing Society*, Nottingham, September 1987, 256–270.

Singh A (1989) Digital change detection techniques using remotely-sensed data. *International Journal of Remote Sensing*, **10** (6) 989–1003.

Stone R J (1986) Radar system and ground parameters–an interactive case study of semiarid terrain in Tunisia, Mapping from Modern Imagery. *Proceedings of a Symposium held by Commission IV of the International Society for Photogrammetry and Remote Sensing*, Edinburgh, Scotland, 8–12 September 1986, 261–270.

Tanre D, Deroo C, Duhaut P, Herman M, Morcrette J J, Perbos J and Deschamps P Y (1990) Description of a computer code to simulate the satellite signal in the solar spectrum: the 5S code. *International Journal of Remote Sensing*, **11**, 659–668.

Townshend J R G and Justice C O (1986) Analysis of the dynamics of African vegetation using the Normalized Difference Vegetation Index. *International Journal of Remote Sensing*, **7** (11) 1435–1445.

Trefois P and Volon C (1992) Lithologic mapping with Landsat Thematic Mapper guided by laboratory spectral measurements, Air, Niger (Africa). *Proceedings of the Central Symposium of the 'International Space Year' Conference*, Munich, Germany, 30 March–4 April 1992, ESA ISY-1, 565–569.

US Geological Survey (1988) Mapping hydrothermally altered rocks with Landsat data. In Denegre (1988).

Vinogradov B V and Frolov D E (1992) Aerospace monitoring for ecological hazard and disaster mitigation. *Proceedings of the Central Symposium of the 'International Space*

Year' Conference, Munich, Germany, 30 March–4 April 1992, ESA ISY-1, 615–618.

Wiegand C L, Gausman H W, Cuellar J A, Gerbermann A H and Richardson A J (1973) Vegetation density as deduced from ERTS-1 MSS response. *Third Earth Resources Technical Satellite Symposium*, **NASA SP-351** (1) 93–116.

Wilkinson G G (1991) The processing and interpretation of remotely-sensed satellite imagery—a current view. In Belward and Valenzuela (eds.) (1991), 71–96.

Wooding M G (1985) SAR image segmentation using digitized field boundaries for crop mapping and monitoring applications. *Microwave Remote Sensing Applied to Vegetation, European Space Agency*, SP-227, 93–98.

Yamagata Y and Akiyama T (1988) Flood damage analysis using multitemporal Landsat Thematic Mapper data. *International Journal of Remote Sensing*, **9** (3) 503–514.

6

Integration of geocoded satellite images in geographic information systems

One of the most significant developments in the handling of geographic information (comparable perhaps to the invention of the map itself) has been the advent of geographic information systems (GISs). GIS technology is evolving rapidly and there is by no means common agreement as to what comprises a GIS, one widely accepted definition being '... a powerful set of tools for collecting, storing, retrieving at will, transforming and displaying spatial data from the real world for a particular set of purposes' (Burrough 1986). This chapter describes how geocoded satellite images can be used as an integrated component of such a GIS and contains the following sections:

- integrated GIS: components of a GIS, the geographic database, a model for an integrated GIS, how time can be incorporated within a GIS and error modelling;
- application examples: satellite images as a GIS source, map-guided image classification and some examples of environmental modelling;
- knowledge-based GIS: the 'next generation' of GIS including automatic extraction of linear features such as roads, knowledge-based image segmentation and the use of neural networks;
- towards a global GIS: global environmental modelling, the earth observing system (EOS) and current problems and prospects.

6.1 INTEGRATED GISs

Until relatively recently, the approach adopted with much remote sensing research was to analyse and interpret images in isolation, that is not making use of any external information. With most images, however, the use of external geographic information (such as digital map data) can assist greatly in the interpretation process, the results of which can in turn be used to update some underlying 'world model'—for example an algorithm designed to extract road networks from SPOT images could use a digital map of existing roads to provide a more focused search space. This is an example of the integrated use of

GISs which is an essential development in realising the potential of satellite images as a geographic information source—a potential that has very rarely been utilised.

6.1.1 Components of a GIS

The components of a typical GIS can be seen in figure 6.1. The heart of any GIS is the geographic database which is described in more detail in section 6.1.2. The other components will vary depending on a particular GIS implementation but can usually be divided into four main functional sub-systems:

— data input;
— data storage and manipulation;
— data analysis;
— data output.

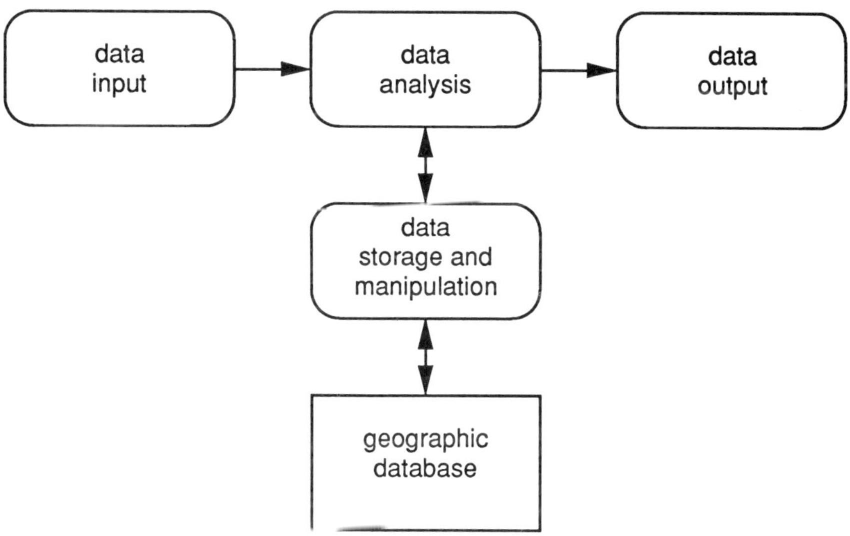

Fig. 6.1. Components of a typical GIS.

Note that this generic system contains aspects typical of many information processing systems—it is the nature of the data and the applications that gives a GIS its unique flavour. The essential features of a GIS can be described by analogy with the more traditional representation of spatially referenced data on a map sheet. This provides three basic sources of geographic information for each feature:

location— by means of accurate cartographic drafting and the use of scales and grids;
description— by means of shading, symbol and annotation;
relationship— by the accurate cartographic positioning of different features relative to each other.

Data input

The data input sub-system is responsible for the capture of a variety of different data formats. Typical categories of input data include:

— map sheets;
— digital map data;
— digital elevation models;
— satellite and aerial images;
— non-spatial or tabulated data, such as statistical or demographic information.

Map sheets can either be digitised by hand (using a digitising tablet) or by automatic scanning. The resultant digital data will be subjected to some form of validation or quality control, such as interactive editing. Digital elevation models can either be input in a standard format or generated internally from digitised contours, before being subjected to a similar validation process. Satellite and aerial data will either be in digital format or may be digitally scanned if the original data is photographic. Tabular data may be input manually, accepted in a suitable digital format or digitised from hard copy using an optical character reader (OCR). The data input sub-system will also be responsible for any necessary format conversion, such as vector to raster or from an external cartographic exchange format to that used internally by the geographic database.

Data storage and manipulation

The data storage and manipulation sub-system is used to maintain the geographic database (see section 6.1.2) and should be capable of retrieving the data necessary to answer a specific user query rapidly and effectively. Typical functions of this sub-system include:

— create a new object in the database;
— retrieve an existing object;
— modify a database object;
— store the object in the database;
— delete a database object.

The data storage and manipulation sub-system must utilise effectively the three previously described primitives (location, attributes and relations) since almost any query posed to the GIS can be decomposed and expressed in terms of these primitives (that is where a feature is, what its characteristics are and how it is related to other objects in the database). The sub-system should also allow a user to browse interactively using a variety of search criteria such as area of interest, data source, or application area. Other functions of the sub-system include the ability to manipulate the data in various ways, for example to convert from one map projection to an other, to rescale data, or to derive secondary information such as deriving slope and aspect maps from a DEM.

Data analysis

The data analysis sub-system is used to combine the information available from the geographic database in order to solve a particular applications problem. It should ideally be capable of answering queries posed in the applications language such as:

— what are the boundary coordinates of this field?
— what is its primary crop?
— how close is it to the nearest major road?

Using queries based on the three geographic primitives more complex applications problems may be framed. Consider for example the situation of a planning officer who wishes to know all the locations in a region suitable for housing development (Williams 1984). The corresponding query might be framed as:

> 'find the location of all areas greater than 10 000 m^2, on land, below 150 metres, on a slope of less than 1:50 and within 3 km of a major road'

One way of solving this problem within a GIS would consist of the following steps (the term 'map' is used as a generic data object without consideration of its actual format):

(1) assemble the data sets required to solve the problem:
— digital map of land areas (D1);
— digital elevation model (D2);
— digital map of roads (D3);
(2) use the digital elevation model (D2) to produce a map of areas below 150 metres (D4);
(3) use the digital elevation model (D2) to produce a map of areas of slope less than 1:50 (D5);
(4) use the map of roads (D3) to produce a buffer-zone map of areas within 3 km of a major road (D6);
(5) combine maps D1, D4, D5 and D6 to produce a map of areas on land, below 150 metres on a slope of less than 1:50 and within 3 km of a major road (D7);
(6) sweep map D7 to remove all areas of less than 10 000 m^2.

One of the powerful aspects of a GIS is the ability to rework the problem with modified parameters (a 'what-if?' exercise). For example the parameters of the above query can be easily modified to solve the problem:

> 'find the location of all areas greater than 50 000 m^2, on land, below 100 m, on a slope of less than 1:25 and within 1 km of a major road'

More advanced uses of a GIS for data modelling can be found in section 6.2.3.

Data output

The data output sub-system presents the result of a GIS session in a variety of ways. For example the result of a query could be a printout of the objects satisfying the constraints of the query. Alternatively, it could be a more detailed statistical summary or a complete vector or raster geographic map. The actual format of the output will depend on the GIS capabilities and hardware configuration but two trends are likely to continue:

(1) the use of high-performance workstations to allow the user to conveniently view and edit the output, and to interact to control the process of a GIS operation;
(2) the ability to interconnect with an external GIS and other environmental databases in an open system.

6.1.2 The geographic database

There are many possible approaches to the data model that underpins the geographic database, with a great deal of current research aimed at providing a suitable object-oriented database, for example, to permit the use of objects as attributes of other objects, such as associating scanned photographs of motorway bridges as attributes of a motorway object stored in vector format. It is still commonplace however for a geographic database to effectively consist of three separate databases, corresponding to the three main data formats:

(1) raster (for example images, scanned maps and DEMs);
(2) vector (for example digital maps, contours and point data);
(3) tabular (for example attribute and statistical data).

Raster data

Raster data is the 'native format' for satellite images and data such as scanned map sheets, and consists of a regular array of picture elements or pixels each of which is referenced by a two-dimensional (column, row) index. The raster format is convenient for inputting large amounts of bulk data but has two potential disadvantages:

(1) it can be wasteful of space when storing scanned maps; this is because digital maps are characteristically more sparse than satellite images and when transformed to raster format will tend to have a lot of blank pixels;
(2) the raster format does not contain topological information about the relative location of geographic objects.

One alternative raster format is the quadtree—this is constructed by dividing an image into quarters, then these quarters into further quarters, and so on hierarchically until the entire image is divided up. The end of any 'branch' on the quadtree is defined when any of the square blocks produced by the splitting process at a particular level contains pixels with all the same value. Thus the quadtree represents the values of the image by a hierarchical tree of homogeneous squares of various sizes (see figure 6.2).

One of the advantages (Hogg and Gahegan 1986) of using quadtrees is that they can be used in a top-down manner to provide a representation of the image at various scales, with all finer detail being hidden at lower levels of resolution. The quadtree representation is extremely efficient for performing arithmetic and deriving image statistics, and is relatively easy to implement in software.

Vector data

The representation of spatial information as vectors is commonplace in most GISs, particularly for digital map data. Vector representation makes use of three simple topological objects:

(1) points (zero-dimensional objects);
(2) lines (one-dimensional objects);
(3) areas (two-dimensional objects).

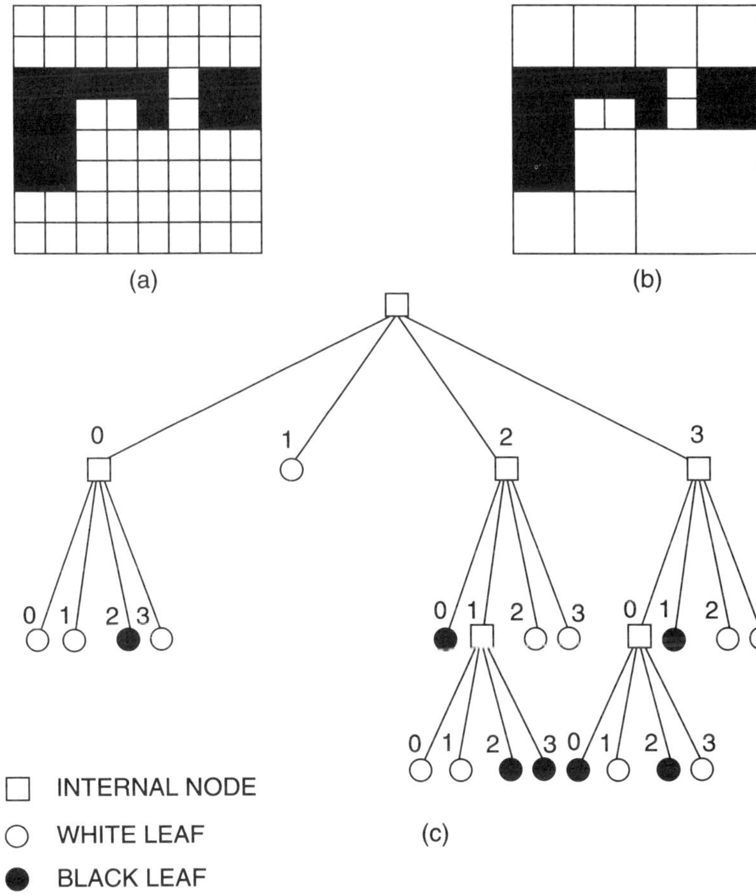

Fig. 6.2. Representation of binary raster data by a quadtree: (a) the original dataset; (b) the leaf quad-square diagram; (c) the resultant quadtree (from Ibbs and Stevens 1988).

The location of a single point can be expressed by its (X, Y) position in some coordinate system. A straight-line segment can similarly be expressed by the location of its two end-points (X_0, Y_0) and (X_1, Y_1), or expressed as vectors $X(0)$ and $Y(0)$ (see figure 6.3).

A curve can be approximated by a series of line segments, for example those determined when an operator digitises a map. A typical curve $X(0), X(1), X(2),\ldots,X(n)$ is shown in figure 6.4.

Finally an area can be approximated by a polygon representing its boundary, the polygon is similar to a curve except that it must be closed: $X(0), X(1), X(2),\ldots,X(n) \rightarrow X(0)$. Another variant of the vector format is the triangulated irregular network (TIN) which is mainly used for the representation of DEMs (Peucker *et al.* 1978). In a TIN a network of triangles is constructed with its vertices at sample points.

As well as the ability to store objects as points, lines or areas it is necessary to store the topology or spatial relationships between objects. For example consider two adjacent

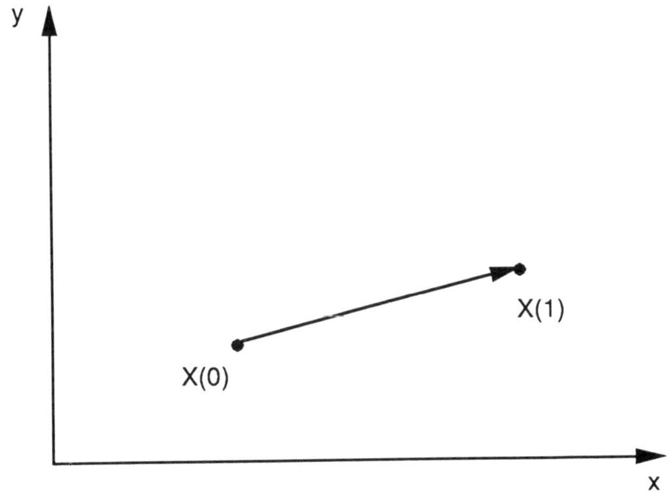

Fig. 6.3. Representation of a straight-line segment in vector format.

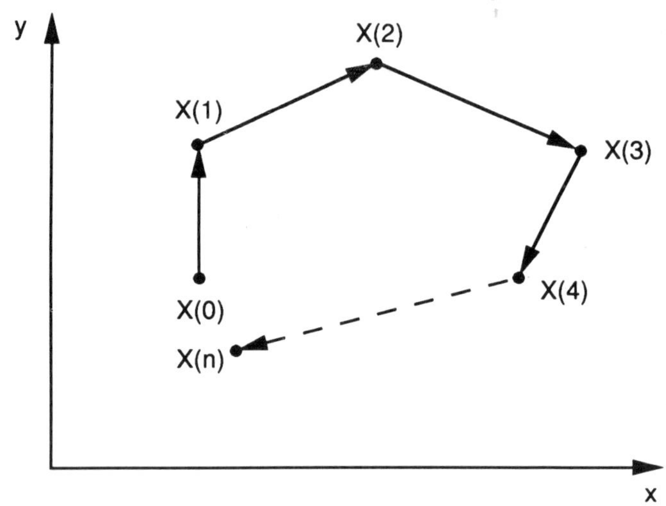

Fig. 6.4. Vector approximation of a curve as a series of line segments.

fields which have been digitised from a map (figure 6.5). The common boundary has been digitised twice, and inevitably there will be some discrepancy in the values recorded; this can give rise to spurious sliver polygons between the two positions of the boundary.

This can be avoided by only digitising the boundary once and indicating that it actually belongs to two separate objects. Other topological aspects that need to be handled include dealing with areas with 'islands' and 'holes' in and polygons that contain loops. Various topological models can be used (Valenzuela 1991). A typical one of these is the

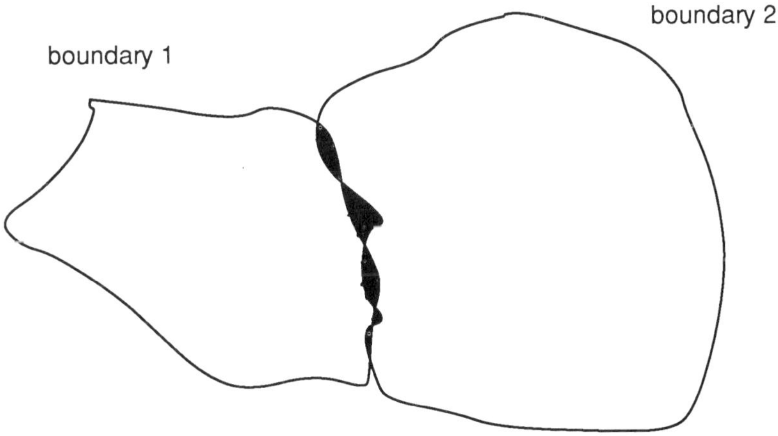

Fig. 6.5. Spurious sliver polygons may be produced by 'double digitising'.

arc-node model where an arc is a sequence of connected vectors which begins and ends in a node which can be the point of intersection between two or more arcs.

Tabular data

Tabular data consists of virtually any data that can be stored in a standard relational database format as a number of records each containing fields with defined values. Thus tabular data could include the population of major cities in an area, traffic flow statistics for a road network or parameters associated with a particular species of tree.

One of the most important types of tabular data within a GIS is attribute data, which is additional information linked to a geographic object to describe its properties, for example the ownership of a field, the materials used to surface a road, or the land-cover classes associated with a classified image. An example of the tabular representation of attribute data (in this case of a forest stand) is shown in table 6.1.

Table 6.1. Tabular representation of attribute data (from Barker 1988)

Operating area number	5
Operating area acres	327
Site index	80
Forest type	pine
Topography	flat
Age	38
Condition class	mature
Volume	32.45

Other classes of attribute data include status and object history information. The status information (Jackson 1988) stored with a geographic object provides additional background data such as area of coverage, date of compilation, geographic accuracy, and applicable standards. Additional status information may include the geographical location, map projection, scale, and feature codes. The object history provides an audit trail of the data that was used to produce the object and the processing performed, tracing back to the original source of the data; for example, a historical record for a DEM might include the identity of the original map containing the topographic information, the process used to digitise the contours and the algorithm used to interpolate contour-heights to a regular grid format. Without such information, little confidence can be placed in the accuracy and reliability of the other attribute data.

6.1.3 An integrated GIS model

A generic model which can be used to represent an integrated GIS is shown in figure 6.6. It has the following four components:

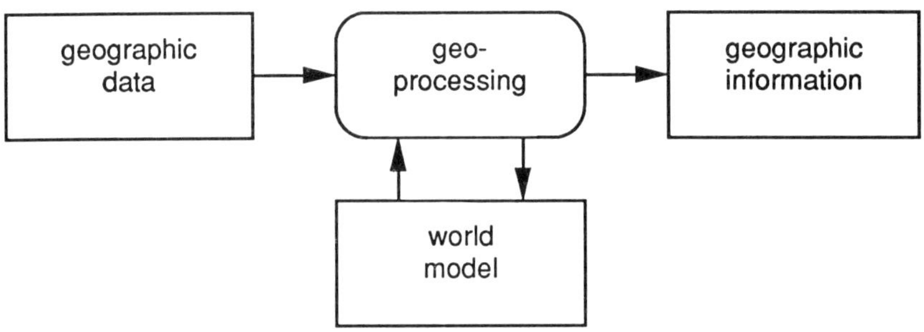

Fig. 6.6. A generic model for an integrated GIS with the following components: (1) geographic data is input to the system; (2) geoprocessing converts the data to useful geographic information; (3) dynamic world model used to influence geoprocessing; (4) geographic information is output from the system.

(1) geographic data is input to the system;
(2) geoprocessing is used to convert the raw data into useful information;
(3) a 'world model' is used to store the geographic information essential for geoprocessing;
(4) geographic information is output from the system.

The world model is the most important component of this representation and has three essential functions:

(1) to constitute a mathematical representation of a fragment of the world at a particular moment in time;
(2) to provide context information used in conjunction with the input data to enhance the geoprocessing;
(3) to provide the capability of being dynamically updated by the geographic information flow.

It is this last that provides the defining characteristic of an integrated GIS using satellite data. The satellite data is used not only as part of the geographic information flow (as happens for example in three-dimensional simulation) but is incorporated in the feedback loop to update the world model. One of the simplest methods for extraction of thematic information from satellite images is the direct digitisation of features within an integrated GIS. The image is geocoded and used as a backdrop to a set of digital map data (for example a road network). The result is very similar to that of an image map. The synergistic advantages of such an approach are readily appreciated. Firstly the display of the digital map data guides the interpreter towards the areas of interest on the image. Secondly it becomes apparent where the map information is out of date, for example where new roads have been constructed. The operator can then draw on the new information as an overlay. The resultant vectors can be edited, manipulated (for example attribute data added) and validated before being included in the geographic database.

Another example of a world model within the context of an integrated GIS might be a DEM derived from a variety of sources. For example a relatively crude DEM may be available based on existing contour maps of an area—this can form the basis of a 'world model' of the heights within a region of interest. This initial model can then be used to provide context information for more accurate elevation data generated from satellite data, for example by optical stereo matching or SAR interferometry. Finally the DEM refinements derived in this stage are used to update the world model—this can be viewed as a dynamic process with the DEM being continually updated with more accurate data every time that new geographic information becomes available.

6.1.4 Temporal evolution

The other factor that must be considered in conjunction with each geographic object is its temporal nature; this requires a world model that can evolve over time (see figure 6.7).

However, even in the (hypothetical) situation that no real world changes had occurred it is still desirable to be able to iterate to a more accurate world model. Such iterations could include the ability to filter out high-frequency noise and to increase confidence in

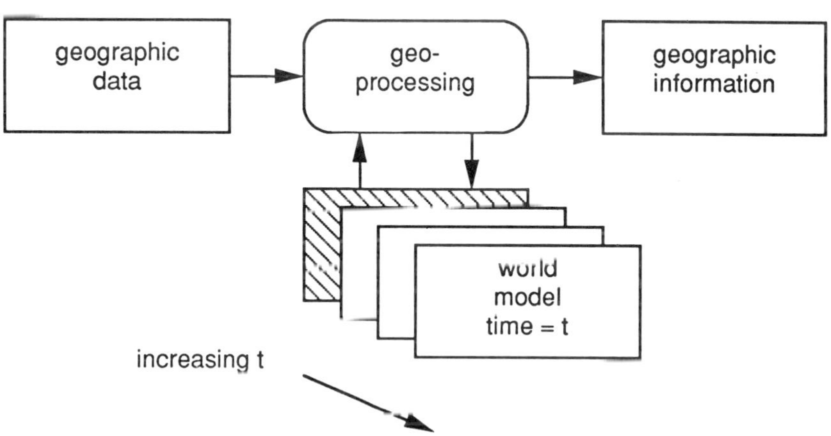

Fig. 6.7. Temporal evolution of the world model.

the stable components of a model as well as the ability to store alternate and historical versions of 'fuzzy' data—one of the characteristics of most complicated mathematical models is that relatively small changes in the starting parameters of a model can lead to very large variations in the state of the model over time. It is important to be able to discern the implications of this for a particular model.

Other aspects of temporal evolution include:

(1) modelling of the changes that have occurred over time (see section 5.4);
(2) the ability to represent objects that are not homogeneous in time, for example map sheets may be revised selectively over a number of years and will never present a 'snap-shot' of an area at a particular moment in time;
(3) the ability to 'roll-forward' the world model to predict the status of geographic objects at some stage in the future, for example the nature of global tropical deforestation in the next decade.

6.1.5 Error modelling

Because the information within a GIS comes from a variety of disparate sources, all with their characteristic errors, it is important to be able to perform some sort of error modelling in order to determine the confidence that can be placed in the output information. The problem is not solely associated with errors in the geocoding and classification of satellite data, although it is essential that error metrics (such as residual geocoding errors and confusion matrices) be included as attributes whenever satellite data is incorporated in an integrated GIS. There are just as likely to be sources of error in a conventional map (Hutchinson 1982). These include:

(1) spatial variation of attributes such as soil colour or vegetation cover within an area marked as homogeneous on a map;
(2) interpolation of point samples (such as soil type) into arbitrary boundaries between features;
(3) variation of individual interpretation of the same sample data (for example analysis of aerial photographs);
(4) generalisation, that is errors introduced as the result of deriving maps at one scale from those at a different scale without resurveying;
(5) temporal non-homogeneity—typically only major features will be amended when a map is revised leaving many relatively minor changes unrecorded.

As well as these intrinsic errors there may well be additional errors introduced as the result of digitisation, for example the production of sliver polygons where the common boundary between two features is digitised twice with slight variations each time. There are five main categories of error in a GIS (Weir 1991):

(1) errors in source data such as data dropout and striping in satellite images, errors in the original source maps, interpolation of point data and temporal errors in the source data;
(2) errors in data input including inaccuracies in GCP selection, errors in training area specification, digitising errors, and incorrect keying of attribute data;

(3) errors in data storage, for example accuracy lost during raster-to-vector conversions (and vice versa) and numerical approximations in computer arithmetic;
(4) errors in data processing such as geocoding and classification errors in satellite images, DEM interpolation and errors propagated by data set combination;
(5) errors in data output, for example the reproduction limitations of output devices.

In an ideal GIS all sources of data would include statistical information about the distribution of errors such as range, mean and standard deviation, or 95% confidence limits. It is also desirable to be able to 'date stamp' error metrics and to quantify the decreasing confidence in data accuracy as the data ages. It is also desirable to have a set of rules which define the error metric associated when a new object is created from two existing ones. Consider for example a straight-line segment, *A*, with length 100 metres. Because of inaccuracies in measurement there will be an error associated with the length (say 1 metre). Using the notation:

$$\text{length (object)} = [L, \delta L] \tag{6.1}$$

where

$$L = \text{estimated length,}$$

$$\delta L = \text{error in length measurement}$$

then length (*A*) = [100, 1].

Using this 'fuzzy' definition of length a few simple rules for defining the error metric associated with various operations on objects can be formulated. For example the rule for estimating the length of two objects placed end-to-end is given by:

$$\text{length (object 1 @ object 2)} = [L_1 + \delta L_1, L_2 + \delta L_2] \tag{6.2}$$

where

$$@ = \text{geometric operation 'place-end-to-end',}$$

$$L_1 = \text{estimated length of object 1,}$$

$$L_2 = \text{estimated length of object 2,}$$

$$\delta L_1 = \text{error in measurement of object 1,}$$

$$\delta L_2 = \text{error in measurement of object 2.}$$

Thus if line *B* has estimated length 200 metres and error in length 3 metres

$$\text{length } (B) = [200, 3]$$

$$\text{length } (A \mathbin{@} B) = [100 + 200, 1 + 3]$$

$$= [300, 4].$$

Unfortunately it is difficult to quantify many sources of error in geographic information and harder still to produce realistic metrics and valid models for error propagation within a system which, by its very nature, involves many complicated manipulations and combinations of widely differing data sources. Error modelling within integrated GISs, although of utmost importance for the widespread acceptance of the technology, is still very much in its infancy; in the mean-time 'all persons using GIS technology ... must be aware of the errors which can occur and where possible take appropriate steps to keep such errors to an acceptable minimum' (Weir 1991).

6.2 APPLICATION EXAMPLES

6.2.1 Satellite images as a GIS source

One of the advantages of using classified images as a source of thematic information in a GIS is their synoptic coverage. For example, a geometrically transformed Landsat-TM image resampled to 25 metres can form the basis of a thematic map covering an area of 185 km × 185 km—it would be considerably more difficult to obtain a map of corresponding scope and granularity by conventional means. One example of this is the UK land cover mapping project (Wyatt and Fuller 1992), two of the motivating factors for the widescale use of satellite images in this project being:

— continuing improvements in the content, accuracy and availability of satellite image data;
— a user community with clearly defined needs for land-surface data at national and larger scales.

The project is intended to be operational in the sense that it is largely automated, sufficiently accurate for end-user requirements and produces verifiable results. The flowlines used for the project can be seen in figure 6.8.

The primary purpose is to produce '... a national environmental GIS which will allow investigation of interactions between land use, climate and environmental impacts as a basis for a future modelling capability' (Wyatt and Fuller 1992). The project intends to produce a baseline inventory for the UK—not necessarily to provide the capability for repetitive monitoring. Suggested applications of the GIS include:

(1) impact of changed land use using regional and global ecological models;
(2) hydrological modelling such as forecast of river-flows and pollutant loading within aquifers;
(3) mapping of ecological habitats and analysis of the relationship between biodiversity and landscape structure;
(4) use in conjunction with DEMs and digital maps to plan cell-phone networks by modelling attenuation of radio signals by different land-cover types.

Some of the drawbacks found in the project are the effects of mixels, misregistration problems, and the differing nomenclatures used for field surveys and image classes. Single season imagery provided inaccurate classifications, therefore classification was performed using both summer and winter imagery of an area. Two validation methods are used: quality control in the processing stage and accuracy assessment in post-processing. A field survey was conducted at 512 sites, using a stratified random selection technique; this showed a classification accuracy of 85–90% for each land parcel.

Landsat data was also an important input into BASIS, a GIS covering the San Francisco Bay Area (Wilson 1986) as it proved to be the most cost-effective method of

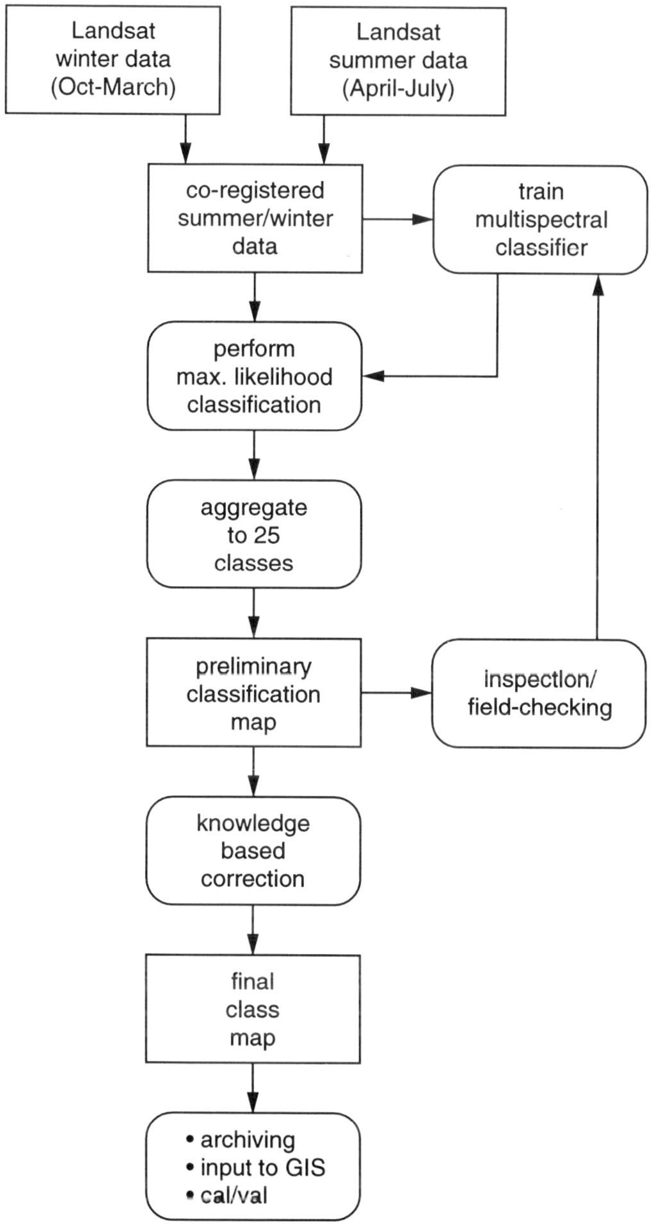

Fig. 6.8. Flowlines used to compile land cover maps of the UK (adapted from Wyatt and Fuller 1992).

collecting regional-scale vegetation and land-cover classes. One application of Landsat data within the GIS was the study of biogenic air pollution. Landsat imagery was classified and used to provide the basic vegetation classification layer. This was then used in conjunction with digital maps of hydrography and urban areas to reclassify categories in marginal areas where there was confusion between urban and non-urban areas that had similar spectral responses. Separate scientific studies were used to determine typical biogenic emission factors for each land cover class (incorporated within BASIS as non-spatial data). This can then be used in conjunction with the other information in the database to estimate total biogenic emission for each grid cell.

Other examples of the use of satellite images as a GIS source include:

(1) environmental sensitivity index (ESI) mapping for oil spills (Jensen *et al.* 1990);
(2) resource management of rice fields in Vietnam (Cuong 1992);
(3) socio-economic analysis of urban areas in Belgium (Donnay 1992);
(4) change in vegetative cover within urban areas of Portland, Oregon (Doak and Lackey 1993);
(5) mapping of wetlands that provide waterfowl habitat in Alaska (Kempka *et al.* 1993);
(6) monitoring of heat islands for micro-climate studies in Singapore (Nichol 1993);
(7) mapping of environmental damage in Estonia (Oberg and Andersson 1993);
(8) a 'seamless' forest cover database of the whole of Canada derived from AVHRR data (Palko *et al.* 1993);
(9) monitoring of environmentally sensitive areas of the UK (Brown *et al.* 1994);
(10) hydrological modelling of lake basins in the Kenyan Rift Valley (Stuttard *et al.* 1994);
(11) maximising of yield and economic benefits of agriculture in Thailand (van Dijk and Galassetti-Morrey 1993).

6.2.2 Classification within an integrated GIS

Another of the advantages of using an integrated GIS model is the additional capability it affords for the classification of satellite imagery, leading to an entirely new category of geographic information product:

> '... classification products produced by integrated GIS are not classified satellite imagery...but are a qualitatively different amalgam combining features of both satellite and cartographic data'.
>
> (Davis and Simonett 1991)

A simple example of integrated classification was used for the analysis of SAR images for agricultural applications (Wooding 1985). The approach was to incorporate digital map data, in this case digitised field boundaries, into the classification process. By using the field boundaries as ground truth, representative statistics such as mean backscatter can be collected for each field and related to the crop type contained therein. DEMs lend themselves particularly to integrated classification since often features of interest are known to be confined to certain elevation ranges or to have spectral characteristics which are strongly correlated with terrain parameters. Such approaches can prove of value in the application of satellite images to vegetation mapping as the topography of the underlying terrain is a major ecological factor in determining the occurrence, distribution and density of a particular species. In the integrated approach to classification (Hutchinson 1982) the sources of information can be introduced at one of three stages of the classification process:

(1) before classification (image stratification);
(2) during classification (classifier modification);
(3) after classification (class sorting).

Image stratification

The idea behind pre-classification stratification is to divide the input image into a series of smaller sub-images, each of which can be classified more easily. This allows, for instance, a division into sub-units each of which is statistically more homogeneous, or the separation of features which are spectrally similar. This is likely to increase classification accuracy because the pixels in a particular class can often have a great spectral variation over an entire image. A typical use of digital data is the classification of soil types, where the areas in an image might be pre-classified using a geological map. The statistical basis of this stratification is that it breaks down the training samples into smaller sets of samples each having a smaller variance than the original data. In many classification algorithms, such as maximum likelihood, the variance is the basis of measuring classification distance; thus with a smaller variance and tighter training clusters greater discrimination is possible. Stratification can also be used to reduce the confusion between spectrally similar features, for example by means of a land-cover map.

Classifier modification

Another approach to incorporation of the geographic information is to use it to modify the classification process itself. The simplest approach is to use the map data as an additional 'band' in the classification. This is a limited technique however due to the differing natures of image and map data. Image data will vary greatly within a single feature class and it is the statistics of this that forms the basis of class discrimination. Digital maps on the other hand describe the earth's surface in terms of a number of discrete classes, for example built-up areas, roads, deciduous woodland. There is no indication of any variance within these classes—thus no useful statistics on which a multispectral classification can be based. This approach can be used in a limited way however with non-parametric classifiers like the box classifier. In this case for a pixel to lie within a particular box it must satisfy two constraints:

(1) it must lie within the training box in feature space;
(2) its map class must correspond to one of the map classes present in the corresponding training sample.

A more sophisticated approach is to use the GIS data to modify the prior probabilities used, for example, in a maximum likelihood classifier. The basic classification algorithm does not make any *a priori* assumptions about the probability of a pixel belonging to any given class; it simply calculates the most likely class based on the sample statistics presented to it. Other geographic information can be used to modify the probabilities—if a forestry map indicates that a pixel corresponds to an area marked as coniferous woodland this provides additional evidence that the pixel should be classed as coniferous.

Note that even if the map is many years out of date it can be used to provide some information: an indication that an area was once coniferous will convey some probability that it still is.

The use of DEMs in classifier modification has certain advantages over other types of digital map data. Firstly, variables such as height and slope can be considered as continuous functions and can thus be more easily integrated into statistically based classification algorithms. Secondly, the derivation of DEMs is objective in its nature—the height and variation of the terrain is a well-defined geophysical quantity—whereas the derivation, for example, of soil and geological maps involves a great deal of subjective division into feature classes and interpolation from sample data by map-makers. A simple example of the use of DEM data as an extra band is to use the DEM as a 'height image' co-registered with the image bands in a box classification. During training the clusters produced show not only the distribution of each class in feature space but also the corresponding height range. Thus the classification algorithm will assign a pixel to a feature class only if it lies within the appropriate intensity range in each spectral band *and* within the corresponding height range. Note that this process is significantly different from stratification—the height ranges are derived from the training samples not from prior information. DEM data can also be used in maximum likelihood classification where the mean height (or slope) and corresponding variances provide additional statistical parameters for determining discrimination surfaces.

Class sorting

Post-classification sorting uses the geographic information to 'reclassify' the output of a classification, for example a single feature class may not be accurately represented by a single well-defined cluster. If the spectral class can be divided into simpler clusters during the classification they can then be merged to form an amalgamated class. This approach can help deal with the problem of feature classes having similar spectral features by breaking down the spectral class into smaller sub-classes then by recombining them according to post-classification rules to produce more accurate feature discrimination. Another approach to post-classification sorting using geographic information is known as 'sorting by evidences' (Veronese and Mather 1993). This technique uses a combination of 'external' evidence provided by topographic information and 'contextual' evidence provided by the spatial arrangement of the classified pixels.

An example of the application of post-classification sorting (Hutchinson 1982) is the approach used for the classification of desert areas in California where there was confusion between various classes such as the bright surfaces of a dry lake bed and strongly illuminated slopes of large sand dunes. In this case the final assignment to appropriate classes was made by applying a series of decision rules based on DEM slope and aspect values (see table 6.2).

Another example of post-classification sorting is the approach used for the production of land-cover maps of the UK (Wyatt and Fuller 1992). This made use of a number of masks (such as urban, upland or coastal) either generated from the classified imagery or from already available cartographic sources for context-based reclassifications including:

Table 6.2. Examples of post-classification sorting rules (based on Hutchinson 1982)

Initial assignment class	Rule	Final assignment class
Active sand dunes	slope < 1%	dry lake
	otherwise	active sand dunes
Stabilised sand	slope < 1%	dry lake
	1% < slope < 15%	stabilised sand
	otherwise	active sand dunes
Low-gradient undissected sandy alluvial fans and washes	slope < 1%	dry lake
	otherwise	low-gradient undissected sandy alluvial fans
Low-gradient cobbly alluvial fans	slope < 8%	low-gradient cobbly alluvial fans
	8% < slope < 15%	dissected cobbly alluvial fans
	otherwise	mountrain scrub
Dissected cobbly alluvial fans	slope < 3%	low-gradient undissected basalt alluvial fans
	3% < slope < 15%	dissected cobbly alluvial fans
	otherwise	mountain scrub
Mountain scrub	slope > 15%	mountain scrub
	otherwise	highly dissected alluvial fans
Shadow and basalt	aspect N or NW	shadow
	slope < 3%	highly dissected alluvial fans
	3% < slope < 8%	slightly dissected basalt alluvial fans
	8% < slope < 15%	dissected basalt alluvial fans
	otherwise	basalt mountains

(1) use of the upland mask to differentiate between characteristic vegetation of the upland and lowland environment;

(2) within the urban mask, areas misclassified as natural land cover are reset to the most likely urban class;

(3) areas classified as maritime vegetation outside the coastal mask are reclassified as the most spectrally similar inland vegetation class.

6.2.3 Environmental modelling

As described in previous chapters a model is a representation of some aspect of reality which can be manipulated according to well-defined, mathematical rules. Essentially there are three main categories of model (Sondheim 1993):

— object models;
— dynamic models;
— functional models.

An object model is a static model of an object's structure and relationships (for example the structure of a satellite sensor in relationship to a satellite platform). This form of modelling represents the real world object directly as a 'snapshot' of its physical form, and is used when geometry and location in relation to other objects is of most importance. A dynamic model represents how an object's state changes with time (for example the position and orientation of a satellite platform) and is used where the timing and sequence of events (the object's history) is of most importance. Finally, functional modelling consists of the use of algorithms (either as explicit equations or using statistical techniques such as regression) which transform sets of inputs to a physical system, for example (Battad and Loh 1993) soil slope and land use to a set of outputs (in this case hydrological response). It is this last approach that is generally of most utility for modelling complicated environmental problems.

The previous examples of satellite imagery as a GIS source and map-guided classification demonstrated a fairly simple use of the world model, essentially a restricted area of the earth's surface represented as a set of grid squares. Each square in the model is represented as a single unique object defined by its cover type. For many environmental applications, however, a more sophisticated mathematical model is required: one capable of defining a number of different processes on varying spatial scales, and quantitatively representing their interaction and temporal evolution. One such example of environmental modelling was a project to investigate the ecological significance of land use change in the United Kingdom (Griffiths and Wooding 1988). Studies of landscape ecology show that the abundance of a particular species is dependent not only on the presence of its preferred habitat within a landscape but also in the spatial arrangement of habitat patches throughout that landscape. Satellite images can be used to derive the land cover for a particular area which can then be used in a more detailed model of the landscape pattern (Baskent 1993) which describes the functional interaction between the various patches that make up a landscape in terms of the flow of energy, material (especially nutrients) and species of plants and animals between them. For example, in a patch-based model of forest stands (Joy *et al.* 1993) a forest is represented as a grid of patches each 0.1 hectares in size. The interaction between patches is defined by a mathematical model which provides parameters for the shading and seed dispersal associated with each patch. The model is then used to predict the possible future evolution of forest patches and their effect on wildlife habitat, forest harvesting and fire disturbance.

Another good example of an integrated approach to an environmental application is an evaluation of the urban environment in Warsaw, Poland (Bochenek and Polawski 1992). The base image used for 1:100 000 scale mapping was acquired from Landsat-TM; after geocoding a digitised urban boundary of Warsaw was overlaid. Other geographic information used included topographic and thematic maps of the Warsaw area, Cosmos space photographs and ground-truth information. The Thematic Mapper image was classified into 11 land-cover/land-use classes which were validated using the supporting geographic information. This classification formed the base layer of the Warsaw database (see figure 6.9); two other thematic layers were also derived from the satellite data:

(1) vegetation classes derived from the normalised difference vegetation index (using bands 3 and 4);
(2) a surface temperature map of the Warsaw urban area derived from the Landsat thermal band (a good application of this under-utilised data set).

On top of the base data, digital soil maps and maps of agricultural use were introduced. These maps included six types of soil and seven types of cropland. The final layer includes statistical information about annual dustfall and sulphur dioxide concentration. Once the GIS database was populated, the next stage was to use environmental models to derive new combinations of data. Three main analyses were performed:

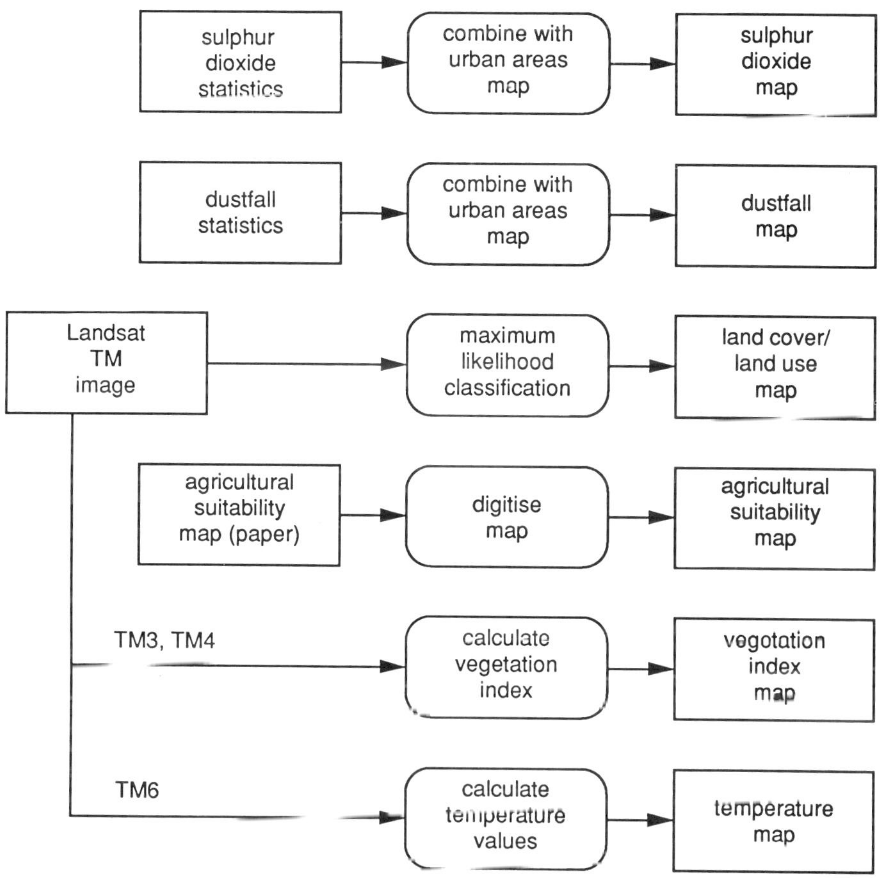

Fig. 6.9(a)

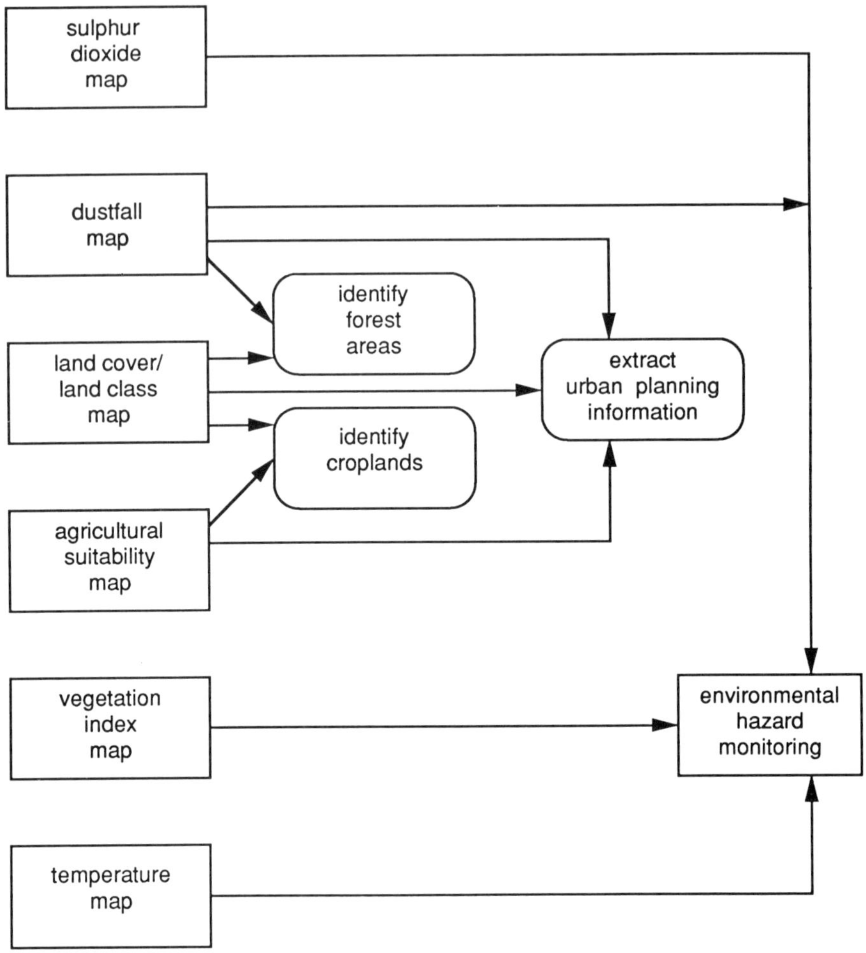

(b)

Fig. 6.9. Integrated GIS used to analyse the urban environment in Warsaw, Poland: (a) production of the geographic database; (b) details of geoprocessing (derived from Bochenek and Polawski 1992).

(1) combination of land use maps and agricultural suitability maps to determine the best cropland categories;
(2) identification of forests within areas of low pollution;
(3) combination of poor agricultural areas, specific land uses and areas of low pollution to determine areas suitable for urban expansion within the Warsaw boundary.

The final stage was to use the urban model to produce a map of environmental hazard within the urban environment. This shows areas that are most suitable for living purposes as well as local 'trouble spots'.

Satellite images, in this case AVHRR, are also an important part of a climate model being used for impact modelling of climatic change in Quebec, Canada (Viau *et al.* 1993). The basic data structure of the model can be seen in figure 6.10 with the three main sources of information being imagery (AVHRR, Landsat and aerial photography), data from climate stations and other terrestrial data such as forest maps and bio-geophysical records.

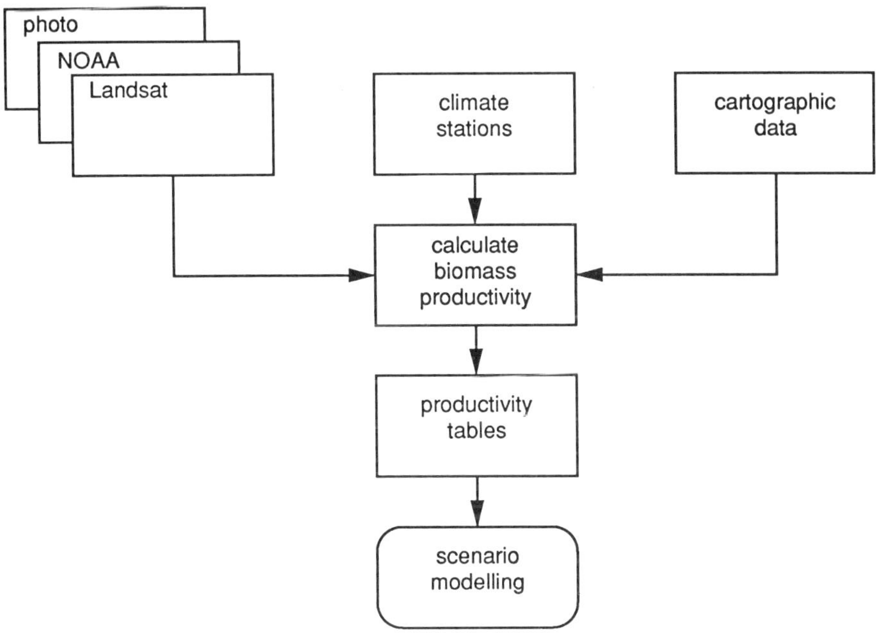

Fig. 6.10. Integrated GIS used for modelling of climatic change in Quebec, Canada (adapted from Viau *et al.* 1993)

The AVHRR data is used to derive a 1 km grid land surface classification of the main ecological domains. This information, together with the other data sources, is fed into the ecological model to produce biomass productivity estimates, which in turn can be used to model the dynamic response of the forest to various climate change scenarios.

The final example of an integrated GIS is provided by a 'Sugar Beet Prediction and Management System' developed by the UK company Logica (Aybet and Walpole 1994). Sugar beet is an economically important crop whose yield can fluctuate greatly due to factors including climatic conditions, seasonal pests and diseases and crop husbandry practices. Sugar companies need accurate and timely predictions to make processing and marketing of sugar more efficient. The integrated GIS approach to this problem utilises a conventional vector-based GIS, an image processing system, a relational database and a numerical predictive model for yield. Multispectral satellite data is collected at three or four dates during the growing season. This is used to calculate leaf-area using a standard NDVI method. The GIS indicated fields where the crop is known to be sugar beet—note that the use of satellite data to monitor growth of a known crop type rather than being used to classify an unknown crop. The NDVI values are fed into the predictive model, along with experimentally determined sugar beet growing constants, and solar radiation derived from a standard UK Meteorological Office product. The final output is a database of estimated yield, in tonnes per hectare, for each factory area being surveyed.

6.3 KNOWLEDGE-BASED GISs

An integrated GIS provides the capability for a skilled operator to include many different sources of geographic data in an integrated analysis to solve a wide range of problems and to generate new geographic information. The next emergent technology in the integrated GIS area is likely to be in the realm of knowledge-based systems which could help automate many of the skilled tasks needed for 'image understanding'. This is particulary important as the number of new earth observation missions planned in the time-frame to circa 2010 will provide a space segment producing quantities of data virtually impossible to analyse by conventional means (even using an integrated GIS). The development of knowledge-based GIS is, however, still very much at the research stage and this section presents three of the active areas of this research:

(1) extraction of roads;
(2) knowledge-based image interpretation;
(3) neural network applications.

6.3.1 Extraction of roads

One area in which knowledge-based techniques have been successfully applied is the extraction of roads from satellite imagery. Not only is the automatic extraction of roads vital for mapping purposes (the rate of building of new roads in most countries rapidly outstripping the map revision cycle) but also for acting as pointers to other human activities (such as urban developments, agriculture and deforestation activities) and the identification of road junctions for use as GCPs. The shape of roads ('long narrow features without many bends') is straightforward to describe mathematically and there is much scope for introducing information about how roads intersect, how they join onto towns and how they appear in conjunction with other image features such as rivers.

The use of feature detection techniques has been discussed in the context of DEM generation (section 4.2.2) and similar techniques can be used to extract the primitive line segments that represent parts of roads. As well as this low-level information a knowledge-based model of how roads 'behave' is used to link the road fragments to form a complete network. Characteristics that can be used to define a road segment include:

— it is a straight line (or has a curvature subject to certain limits);
— it has a known maximum width;
— it has a homogeneous and predictable intensity;
— if multispectral imagery is used then a road will have a characteristic spectral signature;
— it will (generally) have a high contrast with the surroundings;

— roads do not occur above a certain height;
— roads do not occur above a certain slope, or on undulating terrain.

(The last two points suggest the potential benefit of using a DEM.)

Low-level operators can be defined, based on some of the above constraints, to extract candidate road segments from the entire image; however they will also extract other linear features such as canals, rivers and boundaries of fields. A context-based model can be used to help determine genuine road segments. Such a model can make use of two categories of knowledge:

(1) generic knowledge about roads (such as their widths, curvature, the nature of junctions) which can be applied to a wide class of images;
(2) scene-specific information such as the weather conditions and the known positions of existing roads (in terms of the integrated GIS paradigm the 'world model' is a network of roads which can be used to help extract road information from images and which can, in turn, be used to update the knowledge-base).

An example of this approach to road extraction (Ton *et al.* 1989) attempts to classify roads into one of three categories:

(1) major roads—two to four lane highways typically 15–45 metres in width;
(2) local roads—typically 8–20 metres wide;
(3) minor roads.

The low-level road extraction is performed using a series of road-sharpening operators which can extract magnitude and direction of candidate edge pixels. Because roads are characteristically of the order of one pixel wide a road is often surrounded by mixels; by using a deliberately widened ('one-pixel-away') operator there is more likelihood of using pure background pixels. The result is a set of candidate road pixels which are thresholded to form a set of road seed points. A road-following algorithm uses the seed points to produce a 'primitive road file' in which each road segment is classified as level 1, 2 or 3 depending on the result of the road-following algorithm.

The high-level processing consists of three stages:

(1) segment labelling—all level 1 road pixels which are neighbours (in the 8-connected sense) are linked, and all T- and Y-junctions are identified;
(2) segment merging—three parameters are calculated for each segment (length, average direction and average curvature); a series of merge rules (see table 6.3) are applied which allow two segments to be merged;
(3) road labelling— a road is labelled as major if it is long (greater than 100 pixels) and smooth (roughly 10° per pixel), as a local road if it is medium (30–100 pixels) in length and smooth, and minor if it is longer than 10 pixels and has not been placed in one of the preceding categories.

An alternative approach is adopted by Wang and Newkirk (1987) which uses a method similar to unsupervised classification. The preprocessing uses laplacian filters to produce edge images (in each spectral band) which are scaled onto the range [0, 255]. Since only two classes are required ('road' and 'non-road') only two initial cluster points are used for the unsupervised classification. In the case of three image bands the initial points (0, 0, 0) and (255, 255, 255) are used. The result of the iterative process is a binary image where a value of 1 indicates a road.

Table 6.3. Merge rules (based on Ton *et al.* 1989)

Merge two segments IF:

(segments are separated at a junction) AND
(points in directions either side of the junction are consistent)

OR

(segments are disconnected by a very small gap) AND
(all pixels in the gap are level 2 or 3 road pixels) AND
(all directions are consistent)

OR

(segments are disconnected by a small gap [3–8 pixels]) AND
(both segments have similar directions) AND
(length of both segments > three times the length of the gap) AND
(more than half the gap pixels are level 2 or 3 road pixels)

6.3.2 Knowledge-based image interpretation

The traditional approach to multispectral image classification (described in section 5.1) is to process each pixel individually and to attempt to assign it to one of a defined set of classes. In reality there is no reason to suppose that the area represented by a pixel (which is a function of systematic parameters such as sensor IFOV) should correlate exactly with a 'real world' feature. A knowledge-based approach to feature extraction uses models of features, built-up of simple context-based rules, which can then be used to relate the low-level image features (pixels) to high-level real world objects. A simple example of the representation of knowledge in a rule-based system (Yee 1987) is:

IF (side 1 of a road segment is adjacent to water)
AND (side 2 of a road segment is adjacent to water)
THEN (the road segment is a bridge).

Fundamental to the feature extraction process is the ability to segment the image into regions with related attributes. The basic concept is to use a suitable mathematical model to describe real-world objects which can then be extracted using corresponding algorithms. Many segmentation methods are based on edge-detection—objects will appear homogeneous with little difference in reflectivity but will be separated from other object by narrow edges of rapidly changing values.

An example of this is an experimental system developed in the UK for the knowledge-based segmentation of images of land (Corr *et al.* 1989). The input to the system is of two types—a multitemporal sequence of geocoded image data, and a digitised map of the study area. The system works by building an initial model of the image classes, seeded by information from the digital map, and refining it by analysing the multitemporal image set. After initialisation each subsequent image in the sequence is processed using an edge detection filter with an automatically determined threshold. The results are processed to form a network of boundary edge pixels. The system then compares the edge network with the current iteration of the model, by using the edges to assign a confidence to the boundaries currently in the model. Each confidence is a measure of the number of edge pixels within one pixel distance of the model boundary. The regions in the segmentation are then refined using the external knowledge in one of the following ways:

(1) regions may be split into smaller regions;
(2) regions may be merged into larger regions;
(3) a region's class (or confidence level) may be changed.

The initial knowledge from the map data includes classes of linear features (roads, railways, rivers, paths) and areal features (such as fields, forestry, urban areas and inland water). The knowledge-base is used as the basis for a series of segmentation rules. For example one 'split' rule is provided to cater for fields which are joined by a narrow strip and further rules specify classes which are not to be split: these include urban areas, water features and wood. There are also a number of merge rules to amalgamate large homogeneous features from several smaller components; these are based on attributes such as the shape of the common boundary, contrast and length of boundary and the concavity of the region. The consistency rules are used to adjust the confidence in the class based on the available evidence. Map data is incorporated where appropriate: for example a 'water body' rule states that the height contours must not change across a water feature. The confidence in a region is also strengthened by comparison with its neighbour; for example compatible classes are field and field-boundary, urban and road, or where two neighbouring regions have the same class.

Another example of a knowledge-based approach is that used for monitoring of urban areas in Europe (Barnsley 1993). The objective of the project is to use satellite images together with digital mapping to obtain accurate and consistent statistics about urban areas within the EC. Traditional pixel-based approaches are rarely successful on urban areas which present a complex patchwork of land-cover types such as residential and industrial buildings, recreational areas, transport and water features. The methodology adopted is to infer land use statistics from the spatial arrangements of land-cover classifications using a two-stage process:

(1) a low-level, per-pixel classification into a number of broad land-cover types;
(2) 'spatial reclassification' into land use classes based on both the frequency and spatial arrangement of the class labels within each part of the image.

Thus, for example, residential districts may be characterised by the expected spatial arrangement of roads, building materials, grass and trees.

6.3.3 Neural network applications

Traditional multivariate classification methods can become increasingly inapplicable to multisource data due to its being 'multitype' (Benediktsson *et al.* 1990), for example spectral data, elevation range and non-numerical class types. One approach that is becoming increasingly adopted is the use of neural networks, a typical example of which can be seen in figure 6.11.

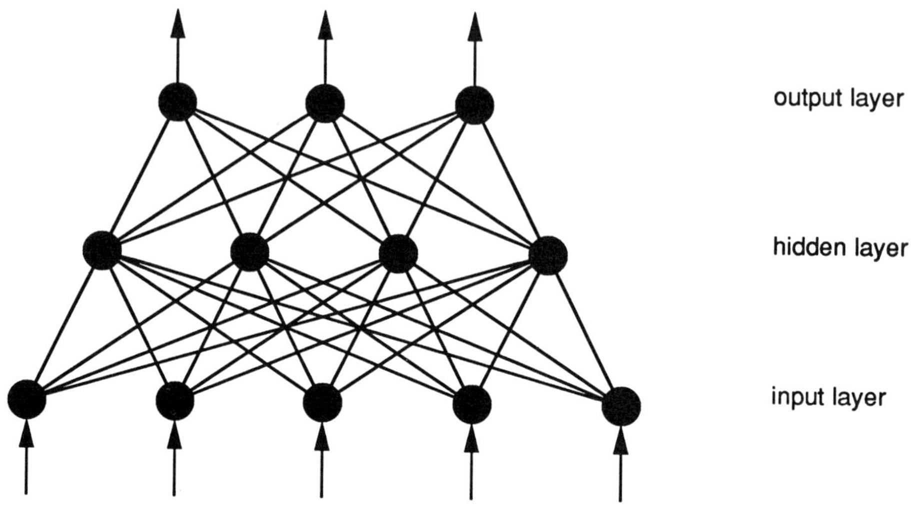

Fig. 6.11. Schematic representation of a typical neural network.

This particular network consists of a number of 'neurons' each of which can have several inputs and only one output. There are a great variety of possible architectures for neural networks and the determination of the optimum configuration for particular applications is currently a very active area of research. The output of a neuron is determined by a function which combines the values of all the inputs connected to that neuron, for example by a weighted sum:

$$F(x) = \sum_i w_i x_i \tag{6.3}$$

where

x_i = value of ith input,

w_i = weighting to be used.

This is often used in conjunction with a threshold value such that the neuron only 'fires' if the set of input stimuli produce an output value above the threshold. Neural networks are wired so that a set of stimuli on the input layer propagates through the internal neural connections ('hidden layers') to produce a second set of signals on the output layer. Thus a neural network can be viewed as a black-box which takes a number of input values (such as grey-levels within an image window) and produces a number of output values (such as probabilities of being within a particular class). In order to train such a network it is presented with a number of samples of different image features together with the class that they represent. During the training process the network continually adjusts the weights of the functions controlling neuron firing until it 'knows' the optimum configuration for predicting the class value of a feature. It is also possible to make use of unsupervised neural networks which divide the data presented into classes based on similarity of patterns, without use of training data, in a manner analogous to that of unsupervised multispectral classification (see section 5.1.3).

The neural network approach has the advantage that no prior knowledge of the statistical distribution of an image is needed. Another advantage of using neural networks is that they can take care of the weighting of evidence from different sources. One of the philosophical objections to the use of neural network techniques is that they operate without a specified mathematical model of the problem in hand (for example feature extraction from a satellite image). Instead a neural network will derive its knowledge of a problem domain from training samples in different categories of interest. This can lead to two main problems:

(1) large numbers of training samples are required to set up the network;
(2) the method may not be robust to features that do not resemble the training data.

The applications of neural networks to classification of images is still at an early stage but indications are (Hepner *et al.* 1990) that their performances are comparable to that obtainable with conventional multispectral classifications and may in some cases be superior. Eventually the optimum approach may be to use neural networks in conjunction with a more modelled knowledge-based approach.

6.4 TOWARDS A GLOBAL GIS

6.4.1 A global environmental model

There is still much controversy about the degree of threat and possible timescales involved in the greenhouse effect on global climatic change. What is clear, however, is that rapid action is necessary to be able to model and quantify changes taking place in the atmosphere ocean–land system of the earth. The problem has been formulated as follows:

> '... we must be able to measure the effect we are having on the planetary environment in a way that allows us to distinguish between the natural evolution of planetary systems and changes which are caused by our activity. To do this requires that we be able to make measurements of the system as a whole, which can only be done efficiently from the vantage point of outer space...' (Macdonald 1992)

Areas where rapid developments are imperative include:

(1) establishing a detailed model of the earth as a system;
(2) increasing the scope and accuracy of terrestrial and atmospheric observations from space;
(3) integration of space and surface observations;
(4) comparison with historical resources to assess evolution of processes on various timescales.

The most important development is to be able to model global climate change in order to quantify the extent, causes and consequences thereof, in particular to be able to monitor and predict average global temperature. It will be necessary to be able to model using a hierarchy of scales from global-scale parameters, through regional scale effects (such as incidence of storms, impact of changes in sea-level and patterns of agriculture) down to local-scale effects. An example of the main components of such a model are summarised in table 6.4.

Table 6.4. Components of a global model (based on Shaffer 1992)

Component	Aspect
Land	land cover human uses of land agriculture/forestry effect on the water cycle
Volcanoes	effect on climate
Ice	arctic/antarctic ice sheets ice dynamics relation to sea-level
Ocean	circulation patterns biological aspects interface with the atmosphere
Meteorology	solar radiation rainfall water vapour clouds
Chemistry	interface between ozone chemistry and the biosphere
Gases	production and removal of greenhouse gases atmospheric transformations

The development of global environmental change models is currently a very active area of earth observation research. For example the UK Natural Environmental Research Council (NERC) has launched the Terrestrial Initiative in Global Research (TIGER) program (Briggs 1992). The TIGER program involves developing models in four key areas:

(1) the carbon cycle on land—with particular focus on the carbon balance of tropical forests;

(2) trace gas exchange between the biosphere and the atmosphere;
(3) global energy and water balance—particularly radiative transfer between vegetative canopies and the atmosphere;
(4) impacts of climate change on various ecosystems.

The use of satellite data to help populate the various models is crucial, for example the carbon cycle model will use multisensor satellite surveys of tropical forest regenerative status as well as global vegetation maps; the impact of climate change on ecosystems will be based on regional scale models of landscape change derived from high-resolution satellite data. Another similar program (Contzen 1992) is the Tropical Ecosystem Environment Observation by Satellites (TREES) project being undertaken by the European Space Agency (ESA) and the Commission of the European Communities (CEC) to assess the scale of global deforestation. The main objectives are:

(1) to provide an inventory of rainforest within the tropical belt;
(2) to determine areas where most deforestation is occurring;
(3) to collect data to help model interactions between tropical forests and the climate.

Optical and thermal data from AVHRR will be used to provide virtually continuous low-resolution monitoring of the earth's tropical belt. Data will be supplemented with higher-resolution observations from Landsat and SPOT over selected areas. Because of the inherent cloudiness of the tropical regions it is envisaged that much use will also be made of SAR data from ERS-1 and ERS-2 missions.

6.4.2 The earth observing system (EOS)

It is clear therefore that to support the rapid development of global environmental models a commensurate development in the quality, quantity and diversity of geographic information from space is necessary. One major initiative in global environmental monitoring is the Earth Observing System (EOS), part of the NASA's 'Mission To Planet Earth' (MTPE) which is planned to involve a large number of instruments collocated on various 'polar platforms', together with a vastly expanded and increased performance ground segment to process the unprecedented volumes of data involved.

The components of EOS are still being finalised but will include contributions from NASA, the European Envisat and Metop missions (Readings and Dubock 1994) and the Japanese Space Agency (NASDA), with the first missions due to be flown before the year 2000. The current (1994) status of the various proposed missions can be seen in table 6.5, with some of the candidate instruments being described in more detail in table 6.6.

The current status of the EOS program can be found in 'The Earth Observer' a bi-monthly magazine published by the EOS Project Science Office, Goddard Space Flight Center, Greenbelt, Maryland.

Such an ambitious satellite program will require a similarly concerted effort back on earth and the EOS Data and Information System (EOSDIS) will be the most complicated ground segment ever undertaken for an earth observation mission. The objectives of EOSDIS (Shaffer 1992) are three-fold:

Table 6.5. EOS missions (based on CEOS 1994 and EOS 1994)

Mission	Agency	Launch date	Duration (years)	Instruments (see table 6.6)	Primary application areas
ADEOS	NASDA	1996	3	AVNIR, ILAS, IMG, NSCAT, OCTS, POLDER, RIS, TOMS	oceanography, atmospheric dynamics, atmospheric chemistry, water/energy cycles
ENVISAT 1	ESA	1998	5	AATSR, ASAR, GOMOS, MERIS, MIPAS, MWR, RA-2, ScaRaB, SCIAMACHY	oceanography, land surface, ice/snow, atmospheric chemistry, atmospheric dynamics, water/energy cycles
EOS-AM 1	NASA	1998	5	ASTER, CERES, MISR, MODIS, MOPITT	atmospheric dynamics, water/energy cycles, atmospheric chemistry, cloud properties, air–land energy/carbon/water exchanges, atmospheric profiles
ADEOS II	NASDA	1999	3	AMSR, DCS, GLI, ILAS-2, POLDER, Sea winds	atmospheric dynamics, water/energy cycles, land surface, oceanography
ALOS	NASDA	2000	5	AVNIR-2, DCS, VSAR	land surface, cartography, disaster monitoring
EOS-AERO 1	NASA	2000	3	SAGE III	atmospheric chemistry, tropospheric/ stratospheric aerosol properties
METOP 1	Eumetsat	2000	5	AATSR, AMSU-A, ASCAT, AVHRR/3, GOMI, HIRS/3, IASI, MHS, MIMR, ScaRaB, SEM	meteorology, climatology

Table 6.5. (continued)

Mission	Agency	Launch date	Duration (years)	Instruments (see table 6.6)	Primary application areas
EOS-PM 1	NASA	2001	5	AIRS, AMSU, CERES, MIMR, MODIS	atmospheric dynamics, water/energy cycles, cloud properties, air/sea fluxes of energy and moisture, sea ice studies
EOS-ALT 1	NASA	2002	5	GLAS, SSALT, TMR	oceanography, geodesy/gravity, land surface, ocean altimetry and circulation, ice sheet mass balance, geological features
EOS-AERO 2	NASA	2003	3	SAGE III	atmospheric chemistry, tropospheric and stratospheric aerosol properties
EOS-CHEM 1	NASA	2003	5	HiRDLS, MLS, TES	atmospheric chemistry, atmospheric dynamics, water/energy cycles
EOS-AM 2	NASA	2003	5	CERES, EOSP, MISR, MODIS, MOPITT	atmospheric dynamics, water/energy cycles, atmospheric chemistry, cloud properties, air–land energy/carbon/water exchanges, atmospheric profiles, vulcanology
METOP 2	Eumetsat	2005	5	TBC	as METOP 1
EOS-AERO 2	NASA	2006	3	TBC	as previous EOS-AERO
EOS-PM 2	NASA	2006	5	TBC	as previous EOS-PM
EOS-ALT 2	NASA	2007	5	TBC	as previous EOS-ALT
EOS-CHEM 2	NASA	2008	5	TBC	as previous EOS-CHEM
EOS-AM 3	NASA	2008	5	TBC	as previous EOS-AM
EOS-AERO 4	NASA	2009	3	TBC	as previous EOS-AERO
EOS-PM 3	NASA	2011	5	TBC	as previous EOS-PM
EOS-AERO 5	NASA	2012	3	TBC	as previous ESO-AERO
EOS-ALT 3	NASA	2012	5	TBC	as previous EOS-ALT
EOS-CHEM 3	NASA	2013	5	TBC	as previous EOS-CHEM

Table 6.6. Proposed EOS instruments (based on CEOS 1994)

Acronym	Full name	Missions—up to and including EOS-AM 2 (see table 6.5)	Description
AATSR	Advanced Along-Track Scanning Radiometer	Envisat, 1, Metop 1	7 band, 1 km resolution, meteorology, climatology, etc.
AIRS	Atmospheric InfraRed Sounder	EOS-PM 1	2300 channels, atmospheric chemistry/dynamics etc.
AMSR	Advanced Microwave Scanning Radiometer	ADEOS II	passive microwave, 8 channels, atmospheric dynamics, ice/snow etc.
AMSU	Advanced Microwave Sounding Unit	Metop 1, EOS-PM 1	15 channel temperature sounder, atmospheric dynamics etc.
ASAR	Advanced Synthetic Aperture Radar	Envisat 1	dual polarisation, C-band, oceanography, ice, land etc.
ASCAT	Advanced Scatterometer	Metop 1	C-band, meteorology, climatology
ASTER	Advanced Spaceborne Thermal Emission and Reflection Radiometer	EOS-AM 1	14 bands, 15 m resolution, land surface, atmospheric dynamics etc.
AVHRR	Advanced Very High Resolution Radiometer	Metop 1	6 bands, 1.1 km resolution, meteorology, oceanography, land etc.
AVNIR	Advanced Visible and Near Infrared Radiometer	ADEOS, ALOS	electronic scanning radiometer, 5 bands, high-resolution land/coastal zone imaging
CERES	Clouds and Earth's Radiant Energy System	EOS-AM 1, EOS-PM 1, EOS-AM 2	broad-band scanning radiometer, 21 km resolution, atmospheric dynamics
DCS	Data Collection System	ADEOS II, ALOS	*in situ* data
EOSP	Earth Observing Scanning Polarimeter	EOS-AM 2	12 bands, atmospheric dynamics etc.

Table 6.6. (continued)

Acronym	Full name	Missions—up to and including EOS-AM 2 (see table 6.5)	Description
GLAS	Geoscience Laser Altimeter System	EOS-ALT 1	geodesy, gravity, ice/snow etc.
GLI	Global Imager	ADEOS II	34 bands, 250 m resolution, oceanography, ocean biology/colour
GOMI	Global Ozone Monitoring Instrument	Metop 1	atmospheric chemistry, ozone
GOMOS	Global Ozone Monitoring by Occultation of Stars	Envisat 1	limb-sounding spectrometer, atmospheric chemistry, climatology
HiRDLS	High Resolution Dynamic Limb Sounder	EOS-CHEM 1	infrared radiometer, 21 channels, atmospheric chemistry, temperature etc.
HIRS	High Resolution InfraRed Sounder	Metop 1	20 channels, meteorology, climatology etc.
IASI	Infrared Atmospheric Sounding Interferometer	Metop 1	atmospheric chemistry, land surface
ILAS	Improved Limb Atmospheric Spectrometer	ADEOS, ADEOS II	spectrometer for atmospheric chemistry applications
IMG	Interferometric Monitor for Greenhouse Gases	ADEOS	atmospheric chemistry applications
MERIS	Medium Resolution Imaging Spectrometer	Envisat 1	15 bands, 300 m resolution, ocean biology/colour, land etc.
MHS	Microwave Humidity Sounder	Metop 1	5 channels, meteorology, atmospheric dynamics
MIMR	Multi-frequency Imaging Microwave Radiometer	Metop 1, EOS-PM 1	6 channels, atmospheric dynamics, oceanography, climatology etc.

Table 6.6. (continued)

Acronym	Full name	Missions—up to and including EOS-AM 2 (see table 6.5)	Description
MIPAS	Michelson Interferometric Passive Atmosphere Sounder	Envisat 1	limb sounding interferometer, atmospheric chemistry, climatology
MISR	Multi-angle Imaging Spectro-Radiometer	EOS-AM 1, EOS-AM 2	4 bands, 240 m resolution, atmospheric dynamics, land etc.
MLS	Microwave Limb Sounder	EOS-CHEM 1	passive radiometer/spectrometer, 5 bands, atmospheric chemistry/dynamics etc.
MODIS	Moderate Resolution Imaging Spectro-radiometer	EOS-AM 1, EOS-PM 1, EOS-AM 2	36 bands, 250 m resolution, atmospheric dynamics, land etc.
MOPITT	Measurements of Pollution in The Troposphere	EOS-AM 1, EOS-AM 2	4 channel correlation spectrometer, atmospheric chemistry
MWR	MicroWave Radiometer	Envisat 1	dual-channel, atmospheric dynamics/humidity
NSCAT	NASA Scatterometer	ADEOS	surface wind speed and direction determination
OCTS	Ocean Color and Temperature Monitor	ADEOS	mechanical scanning radiometer, 12 bands, oceanographic applications
POLDER	Polarization and Directionality of Earth's Reflectances	ADEOS, ADEOS II	passive optical imaging radiometer
RA-2	Radar Altimeter 2	Envisat 1	dual frequency, oceanography, geodesy etc.
RIS	Retro-reflector in Space	ADEOS	laser beam absorption experiment

Table 6.6. (continued)

Acronym	Full name	Missions—up to and including EOS-AM 2 (see table 6.5)	Description
SAGE	Stratospheric and Atmospheric Gas Experiment	EOS-AERO 1, EOS-AERO 2	earth limb scanning grating spectrometer, 9 channels, atmospheric chemistry etc.
ScaRaB	Scanner for Earth's Radiation Budget	Envisat 1, Metop 1	4 bands, 60 km resolution, atmospheric dynamics
SCIAMACHY	Scanning Imaging Adsorption Spectrometer for Atmospheric Cartography	Envisat 1	grating spectrometer, 8 bands, atmospheric chemistry, climatology
SeaWinds	NASA Scatterometer	ADEOS II	oceanography
SEM	Space Environment Monitor	Metop 1	space environment, solar activity
SSALT	Solid State Altimeter	EOS-ALT 1	single-frequency, geodesy, gravity, oceanography etc.
TES	Tropospheric Emission Spectrometer	EOS-CHEM 1	atmospheric chemistry etc.
TMR	TOPEX Microwave Radiometer	EOS-ALT 1	3 bands, atmospheric dynamics, water vapour etc.
TOMS	Total Ozone Mapping Spectrometer	ADEOS	spectrometer operating at UV wavelengths, atmospheric chemistry
VSAR	Versatile Synthetic Aperture Radar	ALOS	land surface, ice/snow

(1) to improve, distribute and archive data products from precursor missions to EOS (mainly extensions to current programs such as Landsat and NOAA/AVHRR);
(2) to process, distribute and archive data from EOS itself;
(3) to control the EOS spacecraft.

Because of the size and complexity of EOSDIS it will be developed in a evolutionary manner with extensive use of prototyping and feedback from users to constrain the final implementation. The first activity for EOSDIS (known as version 0) will be to use existing global data sets as a basis for the geographic database of EOSDIS. The objective of version 0 will be to release useful data sets to global change researchers by converting them to standard formats; these can then be catalogued and validated. Examples of Earth Observation data sets that may be included can be seen in table 6.7.

Table 6.7. Earth observation data sets for EOSDIS (based on Shaffer 1992 and Wilson and Backlund 1992)

Satellite/sensor	Data set
Landsat-MSS/TM	land cover
AVHRR	sea-surface temperature land cover
Defense Meteorological Satellites	water vapour
Total Ozone Mapping Spectrometer (TOMS)	ozone concentration
Upper Atmosphere Research Satellite (UARS)	stratospheric distribution of ozone and trace genes
Radarsat	land/ocean data sets particularly of polar regions
SeaStar	ocean biology/ocean colour

The completed EOSIDS will make use of eight distributed active archive centres (DAACs) located throughout the United States with interfaces to silimar initiatives with other space agencies such as ESA and NASDA. The DAACs currently identified and their data holdings can be seen in table 6.8.

EOS represents a major step forward in geographic information from space, using a systems engineering approach to solve what is essentially a systems problem. The success of the project is by no means assured, however, and major problems such as the sheer volume of data, its non-heterogeneous nature, calibration and standardisation of data, and the political and economic aspects of national access to an international data set present a severe challenge indeed. It is instructive to quote from a speech that Professor John Townshend, one of the early advocates of global data sets, made in 1989:

Table 6.8. EOS distributed active archive centres (based on Shaffer 1992)

Data holding	DAAC location
Land surfaces	EROS Data Center, Sioux Falls, South Dakota
Hydrological cycle and hydrodynamics	Marshall Space Flight Center, Huntsville, Alabama
Sea-ice and polar processes	Alaska SAR Facility, Fairbanks, Alaska
Climate change, atmospheric chemistry	Goddard Space Flight Center, Greenbelt, Maryland
Physical oceanography	Jet Propulsion Laboratory, Pasadena, California
Radiation budge, clouds and aerosols	Langley Research Center, Hampton, Virginia
Cryosphere	National Snow and Ice Data Center, Boulder, Colorado
Non-satellite trace gas flux data	Oak Ridge National Laboratory, Oak Ridge, Tennessee

> '... it is of utmost importance that we have our information systems in place by the mid-1990's, when the inflow of data becomes a torrent. On the basis of present evidence we will not be ready' (Townshend 1989)

Five years later, in 1994, the global environmental problems are more pressing than ever and the need for a global GIS, making maximum use of geographic information from space, is ever more imperative.

6.5 REFERENCES

Aybet J and Walpole R (1994) An integrated GIS approach to the use of earth observation data—sugar beet prediction and management system, Association for Geographic Information, AGI 94, 15–17 November 1994, Birmingham, UK, 15.1.1–15.1.6.

Barker G R (1988) Remote sensing: the unheralded component of Geographic Information Systems. *Photogrammetric Engineering and Remote Sensing*, **54** (2) 195–199.

Barnsley M (1993) Monitoring urban areas in the EC using satellite remote sensing, *GIS Europe*, October 1993, 42–44.

Baskent E Z (1993) Quantifying forest landscape structure for integrated resource planning. *GIS'93 Symposium*, Vancouver, British Columbia, February 1993.

Battad D T and Loh D K (1993) Framework for the integration of Geographic Information Systems with soil erosion simulation models. *GIS'93 Symposium*, Vancouver, British Columbia, February 1993, 353–361.

Belward A S and Valenzuela C R (eds.) (1991) *Remote sensing and Geographical Information Systems for resource management in developing countries.* Kluwer Academic Publishers.

Benediktsson J A, Swain P H and Ersoy O K (1990) Neural network approaches versus statistical methods in classification of multisource remote sensing data. *IEEE Transactions on Geoscience and Remote Sensing*, **28** (4) 540–551.

Bochenek Z and Polawski Z (1992) Use of satellite data and GIS for evaluating urban environment. *Proceedings of Satellite Symposia 1 and 2 : Navigation and Mobile Communications, and Image Processing, GIS and Space-assisted Mapping, from the 'International Space Year' Conference*, Munich, Germany, 30 March–4 April 1992, ESA ISY-2, 377–379.

Briggs S A (1992) Earth observation in the TIGER programme of the UK Natural Environmental Research Council. Proceedings of the Central Symposium of the 'International Space Year' Conference, Munich, Germany, 30 March–4 April 1992, ESA ISY-1, July 1992, 13–17.

Brown R, Slater J and Askew D (1994) Environmental monitoring of protected landscapes—monitoring environmentally sensitive areas using satellite imagery, within GIS, Association for Geographic Information, AGI 94, 15–17 November 1994, Birmingham, UK, 15.2.1–15.2.4.

Burrough P A (1986) *Principles of Geographical Information Systems for land resource assessmen.* Clarendon Press, Oxford, UK.

CEOS (1994) Committee on Earth Observation Satellites (CEOS), Dossier A, September 1994.

Contzen J-P (1992) Earth observation programmes and plans of the Commission of the European Communities (CEC). *Proceedings of the Central Symposium of the 'International Space Year' Conference*, Munich, Germany, 30 March–4 April 1992, ESA ISY-1, July 1992, 1429–1433.

Corr D G, Tailor A M, Cross A, Hogg D C, Lawrence D H, Mason D C and Petrou M (1989) Progress in automatic analysis of multi-temporal remotely-sensed data. *International Journal of Remote Sensing*, **10** (7) 1175–1195.

Cuong P V (1992) Application of remote sensing and GIS—technology to natural resources management: selecting rice fields in Vietnam. *Proceedings of Satellite Symposia 1 and 2 : Navigation and Mobile Communications, and Image Processing, GIS and Space-assisted Mapping, from the 'International Space Year' Conference*, Munich, Germany, 30 March–4 April 1992, ESA ISY-2, 363–368.

Davis F W and Simonett D S (1991) GIS and remote sensing. In Maguire *et al.* (eds.) (1991), 191–213.

van Dijk A and Galassetti-Morrey S (1993) Can GIS help feed the world? *GIS Europe*, November 1993, 1163–1168.

Doak S C and Lackey L (1993) Assessing change in urban natural areas with multi-date satellite imagery. *GIS'93 Symposium*, Vancouver, British Columbia, February 1993, 1163–1168.

Donnay J-P (1992), Remotely sensed data contributes to GIS socioeconomic analysis. *GIS Europe*, December 1992, 38–41.

EOS (1994) The earth observer, EOS Project Science Office, July/August 1994, 6, 4.

Griffiths G H and Wooding M G (1988) Pattern analysis and the ecological interpretation of satellite imagery. *Proceedings of IGARSS '88 Symposium*, Edinburgh, Scotland, 13–16 September 1988, 917–922.

Hepner G F, Logan T, Ritter N and Bryant N (1990) Artificial neural network classification using a minimal training set: comparison to conventional supervised classification. *Photogrammetric Engineering and Remote Sensing*, **56** (4) 469–473.

Hogg J and Gahegan M (1986) Regional analysis using Geographic Information Systems based on linear quadtrees, Mapping from Modern Imagery. *Proceedings of a Symposium held by Commission IV of the International Society for Photogrammetry and Remote Sensing*, Edinburgh, Scotland, 8–12 September 1986, 113–124.

Hutchinson C F (1982) Techniques for combining Landsat and ancillary data for digital classification improvement. *Photogrammetric Engineering and Remote Sensing*, **48** (1) 123–130.

Ibbs T J and Stevens A (1988) Quadtree storage of vector data. *International Journal of Geographical Information Systems*, **2** (1) 43–56.

Jackson M (1988) *Trends in commercial GIS development, presented at Image Processing 1988*, Blenheim Online Publications, Pinner, Middlesex, UK, 201–213.

Jensen J R, Ramsey E W, Holmes J M, Michel J E, Savitsky B and Davis B A (1990) Environmental Sensitivity Index (ESI) mapping for oil spills using remote sensing and Geographic Information System technology. *International Journal of Geographical Information Systems*, **4** (2) 181–201.

Joy M, Klinkenberg B and Cumming S (1993), An architecture for spatial modelling: linking environmental models with remote sensing and GIS. *GIS'93 Symposium*, Vancouver, British Columbia, February 1993, 925–933.

Kempka R G, Reid F A and Altop R C (1993) Developing large regional databases for waterfowl habitat in Alaska using satellite inventory techniques: a case study of the Black River. *GIS'93 Symposium*, Vancouver, British Columbia, February 1993, 581–591.

Macdonald J S (1992) Delivery of information from Earth observation satellites, *Proceedings of Satellite Symposia 1 and 2: Navigation and Mobile Communications, and Image Processing, GIS and Space-assisted Mapping, from the 'International Space Year' Conference*, Munich, Germany, 30 March–4 April 1992, ESA ISY-2, 243–249.

Maguire D J, Goodchild M F and Rhind D W (eds.) (1991) *Geographical Information Systems: principles and applications*, Longman, London, 1991.

Nichol J (1993) Monitoring Singapore's microclimate. *Geo Info Systems*, February 1993, 51–55.

Oberg M and Andersson C (1993) Estonia: mapping environmental damage by satellite. *GIS Europe*, March 1993, 50–52.

Palko S, Lowe J J and Pokrant H T (1993) Canada's new, seamless forest cover data base. *GIS'93 Symposium*, Vancouver, British Columbia, February 1993, 985–990.

Peucker T K, Fowler R J, Little J J and Mark D M (1978) The Triangulated Irregular Network, Proceedings of the ASP Digital Terrain Models (DTM) Symposium. *American Society of Photogrammetry*, Falls Church Virginia, 516–540.

Rast M and Readings C J (1992) The first mission of the ESA Earth Observation Polar Platform. *Proceedings of the Central Symposium of the 'International Space Year'*

Conference, Munich, Germany, 30 March–4 April 1992, ESA ISY-1, July 1992, 1131–1134.

Readings C J and Dubock P A (1994) Envisat-1: Europe's major contribution to earth observation for the late nineties, *ESA Bulletin*, **76**, 15–28.

Shaffer L R (1992) The Earth Observing System. *Proceedings of the Central Symposium of the International Space Year' Conference*, Munich, Germany, 30 March–4 April 1992, ESA ISY-1, July 1992, 945–950.

Sondheim M (1993) Modelling the real world. *GIS'93 Symposium*, Vancouver, British Columbia, February 1993, 1099–1111.

Stuttard M J, Hayball J B, Narciso G, Suppo M and Oroda A (1994) Use of GIS to assist hydrological modelling of lake basins in the Kenyan Rift Valley, Association for Geographic Information, AGI 94, 15–17 November 1994, Birmingham, UK, 13.4.1–13.4.8.

Ton J, Jain A K, Enslin W R and Hudson W D (1989) Automatic road identification and labeling in Landsat 4 TM images. *Photogrammetria (PRS)*, **43**, 257–276.

Townshend J R G (1989) Mapping, monitoring and forecasting of global/continental change. *Survey and Mapping '89*, The Third UK National Land Surveying and Mapping Conference and Exhibition, University of Warwick, 17–21 April 1989, D2, 1–9.

Valenzuela C R (1991) Spatial databases. In Belward and Valenzuela (eds.) (1991), 311–333.

Veronese V F and Mather P M (1993) Supervised filtering: a postclassification technique based on contextual and ancillary information. *Proceedings of the 19th Annual Conference of the Remote Sensing Society*, Chester College, 16–17 September 1993, 120–127.

Viau A A, Royer A, Boivin J, Ansseau C and Theriault A (1993) Integration of remote sensing data in a Geographic Information System for impact modelling of long-term climatic change in Quebec's boreal forest. *GIS'93 Symposium*, Vancouver, British Columbia, February 1993, 493–497.

Wang F and Newkirk R (1987) A knowledge based system for highway network extraction. *Proceedings of IGARSS '87 Symposium*, Ann Arbor, Michigan, 18–21 May 1987, 343–347.

Weir M J C (1991) Errors in Geographic Information Systems. In Belward and Valenzuela (eds.) (1991), 349–355.

Williams J M (1984) *REGIS—an RAE experimental image-based Geographic Information System*. Royal Aircraft Establishment (Farnborough, UK), Technical Report 84103.

Wilson G S and Backlund P W (1992) Mission to planet earth. *Photogrammetric Engineering and Remote Sensing*, **58** (8) 1133–1135.

Wilson P M (1986) Using Landsat data within a Geographic Information System, Mapping from Modern Imagery. *Proceedings of a Symposium held by Commission IV of the International Society for Photogrammetry and Remote Sensing*, Edinburgh, Scotland, 8–12 September 1986, 491–499.

Wooding M G (1985) SAR image segmentation using digitized field boundaries for crop mapping and monitoring applications. *Microwave Remote Sensing Applied to Vegetation, European Space Agency*, SP-227, 93–98.

Wyatt B K and Fuller R M (1992) European applications of space-borne earth observation for land cover mapping. Proceedings of the Central Symposium of the 'International Space Year' Conference, Munich, Germany, 30 March–4 April 1992, ESA ISY-1, July 1992, 655–659.

Yee B (1987) An expert system for planimetric feature extraction. Proceeding of IGARSS '87 Symposium, Ann Arbor, 18–21 May 1987, 321–325.

Glossary of terms

AATSR
Advanced Along-Track Scanning Radiometer (Envisat)

active sensor
a sensor (such as SAR) that generates its own energy source

affine transform
a simple linear function used to model the geometric transformation of an image (e.g. for **bulk geocoding**)

AIRS
Atmospheric InfraRed Sounder (EOS)

AMSR
Advanced Microwave Scanning Radiometer (ADEOS)

AMSU
Advanced Microwave Sounding Unit (EOS)

ASAR
Advanced Synthetic Aperture Radar (Envisat)

aspect
direction of greatest slope at a location in a **digital elevation model**

ASTER
Advanced Spaceborne Thermal Emission and Reflection (EOS)

atmospheric window
a **spectral band** in which atmospheric attenuation of the electromagnetic signal is small thus making passive remote sensing possible

ATSR
Along-Track Scanning Radiometer (ERS-1)

attitude
the three-dimensional vector which describes the direction that a spacecraft platform is pointing in

attribute data
additional information linked to a geographic object to describe its properties, for example the ownership of a field

autolinear
automatically derived linear **contrast stretch** designed to map the original **dynamic range** (or a designated percentage thereof) onto the maximum available dynamic range

AVHRR
Advanced Very High Resolution Radiometer

AVNIR
Advanced Visible and Near Infrared Radiometer (ADEOS)

azimuth
angle of pointing of a satellite or instrument measured from a fixed point in the plane of the horizon

backscatter
the proportion of the energy incident on a target re-radiated towards an active microwave instrument

band
see **spectral band**

band fabrication
the creation of a set of artificial image bands from pre-existing ones

band ratio
the division of **pixels** in one **spectral band** by those in a second to try to eliminate gross radiometric errors, e.g. sun-angle effects, or as the basis of a **vegetation index**

binary image
an image having only two possible pixel values, often used as an overlay for **masking**

blending function
a function comprised of a linear combination of two component functions—typically used to provide a smooth join between two adjacent sections of an image **mosaic**

blunder detection
method of **validation** based on the trapping of gross errors or physically impossible results

box classifier
a simple algorithm used for **supervised classification** with class values being determined by the range within each spectral band of a **training sample** (also known as parallelepiped classifier)

brightness
the overall intensity (i.e. total reflected energy) of an image **pixel**, often contained in the first component of an image transform such as the **IHS transform** or **tasselled cap** transform

bulk geocoding
the use of simple models of the satellite imaging process to produce roughly accurate geocoded images.

calibration
conversion of data values from arbitrary engineering units (such as pixel values) to meaningful geophysical quantities such as temperature

CCD
Charge Coupled Device

CEC
Commission of the European Communities

CEOS
Committee on Earth Observation Satellites

CERES
Clouds and Earth's Radiant Energy System (EOS)

change image
an image representing the change between two members of a multitemporal image set obtained, for example, by subtracting the oldest from the youngest

channel
alternative name for **band, spectral band**

chip
a small extract taken from a reference image used for the automatic relocation of **ground control points** for subsequent use in **geocoding**

classification
the use of statistical techniques to segment images into different feature classes, usually based on multispectral characteristics

classifier modification
the introduction of additional geographic information during an image **classification**

class sorting
the introduction of additional geographic information after an image **classification**

clinometry
an experimental technique used to derive height information from a single image based on local grey-scale variations (also known as shape from shading)

clipping
calculating where map **vectors** cross an image boundary

cluster
the grouping together of points on a **scatter diagram** (or in **feature space**) characteristic of the response from a single feature class

cluster analysis
see **unsupervised classification**

colour gamut
the range of colours available in a particular medium, e.g. printing or visual display

compositing
see **temporal compositing**

confusion matrix
a method used to validate a multispectral **classification** by comparison of the class values of a sample of points against their actual classes as determined from ground data

contrast stretch
the process of modifying the histogram of an image, usually to enhance visual appearance

convolution
filtering of an image by multiplying the neighbours of each pixel by the weights in a kernel and summing the results

CORINE
CoORdination of INformation on the Environment

cubic convolution
a **resampling** algorithm based on a cubic approximation of the ideal interpolator (sinc function)

DAAC
Distributed Active Archive Centre (EOSDIS)

data compression
any algorithm for encoding large data sets (typically images) so that less computer storage is required. If the original data set can be exactly reconstituted (decompressed) the algorithm is known as 'lossless' compression

DCS
Data Collection System (ADEOS)

density slicing
an image enhancement algorithm which operates by dividing the image histogram into a number of non-overlapping ranges and allocating a colour to each; can also be used as a very simple method of image **classification**

destriping
removal of regular stripe patterns in an image caused by miscalibrations within arrays of sensors, such patterns may be aligned vertically, horizontally (or occasionally both)

DFT
Discrete Fourier Transform

difference of gaussian (DoG)
a non-linear filter thought to be a good approximation of the 'edge detector' employed in the human visual system

digital elevation model (DEM)
representation of the topography of the earth's surface by an array of numbers representing heights above a reference datam

digital map
any representation of a map in numerical format, for example storing line-segments as digital **vectors**

digital parallax model (DPM)
an intermediate model used in the production of a **digital elevation model**. The DPM represents the apparent shift (parallax) in position at a location viewed from two different angles. This can then be converted to the actual height value at the location

digital terrain model (DTM)
often used as a synonym for **digital elevation model**, a DTM may have additional thematic information, such as land-cover classes, as well as height information

digitisation
the process of converting information to digital format, specifically the conversion of paper map sheets to **vector** or **raster** format

digitising tablet
electronic device for manual **digitisation** of paper maps using a mouse or puck equipped with crosswires, which is moved across the map attached to the surface of the tablet

direct image referencing
see **georeferencing**

doppler shift
the change in frequency of a signal (typically radar) due to the motion of the transmitter or the **target**

DRA
Defence Research Agency (Farnborough, UK)

dynamic model
a model which represents how an object's state changes with time

dynamic range
the range of digital (signal) values present in an image

easting
the horizontal '*x*' direction in a rectangular map coordinate system

elevation
(1) height of a point on the earth's surface above a fixed datum (as in a **digital elevation model**)
(2) angle of pointing of a satellite or above (or below) the plane of the horizon

EOS
Earth Observing System

EOSDIS
EOS Data and Information System

EOSP
Earth Observing Scanning Polarimeter

ephemeris
the continuous set of orbital data describing the position and **attitude** of the satellite platform at any given time

epipolar
image geometry meaning that, given a point in the first image of a stereo pair, the match point must lie on a known line in the second image

ESA
European Space Agency

ESI
Environmental Sensitivity Index

EVI
Environmental Vegetation Index

false colour
the selection of a set of image bands whereby terrestrial features are not portrayed in their natural colours

feature space
the *n*-dimensional equivalent of a **scatter diagram** where each dimension represents the pixel value in a particular **spectral band**

filler data
data used to pad-out non-rectangular images so that they may be stored more conveniently

foreshortening
the effect (in SAR images) of points on slopes facing towards the transmitter appearing closer together

functional model
a model based on the use of algorithms which transform sets of inputs to a physical system to a set of outputs

GAC
Global Area Coverage (AVHRR)

geocoding
production of an image where pixel values have been **resampled** onto a regular grid in a standard map projection

geographic database
the component of a GIS where the geographic information is actually stored

geographic information system (GIS)
a system for capturing, storing, checking, integrating, manipulating, analysing and displaying data (such as geocoded images) which are spatially referenced to the earth

geometric correction
obsolete term for **geometric transformation/geocoding**

geometric transformation
(1) the process of creating a geocoded image (synonym for **geocoding**)
(2) the matrix used to define the geocoding function

geoprocessing
the processing of geographically referenced data

georeferencing
producing an image where the transformation to map coordinates is known but may not actually be applied to the image data itself (also known as direct image referencing)

geostationary
a satellite whose orbit allows it to remain above a fixed location on the earth's surface (strictly such a satellite is geosynchronous as its orbit will generally deviate slightly from a fixed point)

GLAS
Geoscience Laser Altimeter System (EOS)

GLI
GLobal Imager (ADEOS)

GOMI
Global Ozone Monitoring Instrument (Metop)

GOMOS
Global Ozone Monitoring by Occultation of Stars (Envisat)

GPS
Global Positioning System

ground control point (GCP)
a point whose position is accurately known on an image and in map coordinates

ground segment
the part of an Earth Observation mission comprising data reception, processing, archiving and distribution facilities, as opposed to the **space segment**

HiRDLS
High Resolution Dynamic Limb Sounder (EOS)

HIRS
High resolution InfraRed Sounder (Metop)

histogram equalisation
a form of **contrast stretch** where the histogram of the original image is adjusted (equalised) to have a uniform distribution

histogram matching
a form of **contrast stretch** where the histogram of one image is modified to correspond to that of a reference image (used extensively for mosaic production)

homogeneous
referring to an image area having little variation in value between neighbouring pixels (i.e. lacking in **texture**)

Hough transform
an image transform used to detect features whose shape can be expressed as parameterised equations such as straight lines and circles

HRV
High Resolution Visible (SPOT)

IASI
Infrared Atmospheric Sounding Interferometer (Metop)

IHS
Intensity Hue Saturation

IIPF
Interactive Image Processing Facility

ILAS
Improved Limb Atmospheric Spectrometer (ADEOS)

image
a data structure storing intensity or other values as a regular array of **pixels** (also known as a **raster**, or raster image)

image map
a product generated from a geocoded satellite image together with additional cartographic information

image processing
any processing (such as contrast stretch, edge enhancement) performed on data in image formats

image registration
an 'unmodelled' alternative to **geocoding** where no map information is available. The two images are directly overlaid by the use of pairs of **tie points**

image repair
substitution of new data values for missing or inaccurate data in an image, one example being **destriping**

image understanding
the use of knowledge based and other advanced tools to obtain detailed semantic information about the real-world scene portrayed in an image

IMG
Interferometric Monitor for Greenhouse Gases (ADEOS)

intensityhuesaturation (IHS) transform
An algorithm used to convert image bands from multispectral (RGB) space to the more intuitive IHS space. The intensity (I) component is related to the overall brightness of a pixel, hue (H) to its colour and saturation (S) to its purity

interferometry
a method of obtaining height information, based on the difference in phase of two SAR signals originating from two slightly different positions

intermediate field-of-view (IFOV)
the portion of the earth's surface that can be seen at any one instant by a satellite instrument

interpolation
(1) any one of a number of algorithms used for image resampling, e.g. **nearest neighbour, cubic convolution**
(2) the estimation of intermediate values from known data points

intervisibility analysis
the determination of which areas can be seen from a given point

IR
Infra-Red

iso-doppler lines
points of the earth's surface with constant doppler frequency for a particular SAR image

jitter
high-frequency **attitude** variation in the orbit of a satellite

kernel
the matrix of numbers used to define a filtering (**convolution**) operation on an image

LAC
Local Area Coverage (AVHRR)

LAI
Leaf-Area Index

layover
extreme case of **foreshortening** in SAR images when the terrain slope is greater than that of the incident signal

linear interpolation
estimation of values intermediate to two data points by using a straight-line approximation to the function being interpolated

line spread function (LSF)
one-dimensional analogue of a **point spread function**

LISS
Linear Imaging Self-scanned Sensor (IRS)

look-up table (LUT)
a table of numbers specifying a transfer function to convert input data values to required output values, e.g. as defined in a **contrast stretch**

masking
use of a binary image to define which parts of an image are to be included in an **image map** or **mosaic** (for example to remove areas of sea or cloud)

maximum likelihood classifier
algorithm used for **supervised classification** with class values being determined by the probability that a point lies within the classes defined by a **training sample**

MERIS
MEdium Resolution Imaging Spectrometer (Envisat)

MHS
Microwave Humidity Sounder (Metop)

mid-infrared (MIR)
the portion of the electromagnetic spectrum between 3.5 and 5.0 μm available for remote sensing

MIMR
Multi-frequency Imaging Microwave Radiometer (EOS)

minimum distance classifier
algorithm used for **supervised classification** with class values being determined by the nearest class centroid defined by the **training sample**

MIPAS
Michelson Interferometric Passive Atmosphere Sounder (Envisat)

MISR
Multi-angle Imaging Spectro-Radiometer (EOS)

misregistration
the misalignment of supposedly coincident points in an multitemporal image set caused by inaccuracies in the geocoding process

mixel
a pixel corresponding to an area on the ground with a mixture of land cover type

MLS
Microwave Limb Sounder (EOS)

MODIS
MODerate resolution Image Spectro-radiometer (EOS)

MOPITT
Measurements of Pollution In The Troposphere (EOS)

mosaic
a combination of two or more geocoding images to form a larger image product

MSS
Multi-Spectral Scanner (Landsat)

MTPE
Mission To Planet Earth

multitemporal analysis
analysis of an area using collocated images acquired at different times. Multitemporal analysis is typically classed as change detection or **multitemporal classification**

multitemporal classification
The inclusion of images from more than one date in a classification set in order to achieve more accurate results, for example by taking into account differences in spectral response during a crop's growing cycle

MWR
MicroWave Radiometer (Envisat)

nadir
referring to an instrument pointing directly downwards from a satellite platform

NASA
National Aeronautics and Space Administration (USA)

NASDA
NAtional Space Development Agency (Japan)

navigation
sometimes used as a synonym for **geocoding**

NDVI
Normalised Difference Vegetation Index

nearest neighbour
a simple **resampling** algorithm based on defining the resampled value to be that of the nearest pixel in the original image

near-infra-red (NIR)
the portion of the electromagnetic spectrum between 1.0 and 2.4 μm available for remote sensing

NERC
Natural Environmental Research Council (UK)

neural network
a non-algorithmic computational paradigm consisting of a number of interconnected 'neurons'. The output of a neuron is determined by a function which combines the values of all the inputs connected to that neuron

NOAA
National Oceanographic and Atmospheric Administration (US)

northing
the vertical '*y*' direction in a rectangular map coordinate system

NSCAT
NASA SCATterometer (ADEOS)

object model
a static model of an object's structure and relationships

OCR
Optical Character Reader

OCTS
Ocean Colour and Temperature Monitor (ADEOS)

OIS
Orbital Imaging System

OMI
Optical Mapping Instrument

orthoimage
a precision geocoded image fully corrected for terrain effects

PA
PAnchromatic (SPOT)

PAF
Processing and Archive Facility (ERS)

panchromatic
an image having a single **spectral band**

parallelepiped classifier
see **box classifier**

passive sensor
a sensor that uses an external energy source, such as the sun

percentile
in a (cumulative) image histogram the value below which a certain percentage of the image values lie, e.g. the 90th percentile is the value below which 90% of the image values lie (the 50th percentile is otherwise known as the median and the 25th and 75th percentiles the lower-quartile and upper-quartile respectively)

photogrammetry
the use of photography for surveying, specifically the measurement of elevations on the earth's surface

piecewise linear
a function composed of a number of straight-line segments (i.e. function is linear between pairs of defined points) often used to provide a customised **contrast stretch**

pixel
a picture element— the smallest component of an image. Also referred to (particularly for SAR images) as a resolution cell

pixel size
the terrestrial area represented by a single pixel in a satellite image

point spread function (PSF)
a two-dimensional function characterising the response of a sensor to the area surrounding the point that it is centred on

POLDER
POLarization and Directionality of Earth's Reflectances (ADEOS)

precision geocoding
geocoding using detailed models of satellite geometry and orbit, **digital elevation models** etc. to produce an accurate transformation of an image to a map projection

preprocessing
any preliminary processing such as **contrast stretch, image repair** performed on the radiometric values of an image prior to geocoding and subsequent product generation

principal components transform (PCT)
an algorithm which produces a decorrelated linear combination of the original image bands

process validation
a method of **validation** using the mathematical model used for a particular process to also generate quality control parameters as an intrinsic part of the process

product validation
a method of **validation** based on quality control of the final output product against expected accuracy standards

proximity map
an image where the values represent the distance of each grid point from features of interest such as roads or urban areas

pseudo-colour
the mapping of the grey-level values of a panchromatic image onto specific colours for presentation purposes. Often used in conjunction with **density slicing**

PVI
Perpendicular Vegetation Index

quadtree
an alternative method for storage of raster images, based on the successive division of the image into quarters until a hierarchy of **homogeneous** blocks is created

query language
a form of interactive computer language used to ask questions (queries) of a relational database

RA
Radar Altimeter

radarmap
alternative term for an **image map** or **mosaic** based on radar data

radiometric correction
the alteration of image values to account for sun-angle, atmospheric attenuation and other illumination effects

range
the distance between a remote sensing instrument (such as SAR) and its terrestrial target

raster, raster image
see **image**

RBV
Return Beam Vidicon (Landsat)

repeat cycle
the time taken by an orbiting satellite between revisits to a particular location

resampling
the process of deriving suitable values for a geocoded image pixel from the original image data values

residual vector
a measure of the accuracy of the **geocoding** process defined as the difference between a GCP's true image position and that given by the calculated transformation

resolution
the size of target that a particular sensor is able to discriminate (which may be variable, depending on target characteristics, contrast with background etc.). Often used as a synonym for **pixel size** or **intermediate field-of-view**

resolution cell
see **pixel**

RGB
Red Green Blue

RIS
Retro-reflector In Space (ADEOS)

RMS
Root Mean Square (error)

SA
Selective Availability (GPS)

SAGE
Stratospheric and Atmospheric Gas Experiment (EOS)

SAR
Synthetic Aperture Radar

saturation
(1) when the actual intensity associated with a pixel value exceeds the **dynamic range** of the sensor or processing system
(2) the 'purity' component of an **IHS transform**

ScaRaB
Scanner for earth's Radiation Budget (Envisat)

scatter diagram
a two-dimensional plot of values in an image where the x-axis represents all the values in band 1 and the y-axis represents the corresponding values in band 2

SCIAMACHY
Scanning Imaging Absorption Spectrometer for Atmospheric Cartography (Envisat)

seed point
point used as the start for various search algorithms, such as the location of a **ground control point**

segmentation
division of an image into discrete 'objects' on the basis of their characteristics in an image (e.g. colour, texture, edges) or by using external information (map-guided segmentation)

SEM
Space Environment Monitor (Metop)

shadowing
effect apparent in SAR images when the local slope of the terrain prevents certain areas from being imaged

shape from shading
see **clinometry**

sinc function
the ideal mathematical interpolation function

SIR
Shuttle Imaging Radar

slant range
SAR image format in which one axis represents the distance from the satellite position to the target, whilst the orthogonal axis represents the **doppler shift**

sliver polygon
a spurious polygon obtained when the common boundary between two areas on a map is digitised twice

slope
magnitude of the local gradient at a location in a **digital elevation model**

Space Oblique Mercator (SOM)
a map projection defined in terms of the satellite position and direction when an image was acquired. The SOM is a conformal mapping along the ground track of the satellite and uses as axes the along-track and across-track directions

space segment
the part of an Earth Observation mission comprising the satellite platform and instruments, as opposed to the **ground segment**

speckle
coherent noise caused by phase variations in a SAR image

spectral band
a specific range of wavelengths within the electromagnetic spectrum

spectral signature
the characteristic amount of incident energy reflected by a particular surface type as a function of wavelength—used as the basis of multispectral **classification**

specular reflection
perfect 'mirror-like' reflection of radar signals from smooth surfaces such as lakes. Because little of the incident energy is scattered back towards the receiver such features will appear dark in a SAR image

SPOT
Satellite Pour Observation de la Terre

SSALT
Solid State ALTimeter (EOS)

stereo matching
any algorithm or technique used to obtain height information from the parallax information contained within two images acquired at different angles (a stereo pair)

stereo pair
see **stereo matching**

stereoplotter
optical device for the plotting of contours from a photographic stereo pair

stratification
the introduction of additional geographic information prior to an image classification

sun-synchronous
a satellite orbit which crosses the equator at the same local time each day

target
a terrestrial feature present in an image, usually in the context of radar

tasselled cap
an algorithm similar to the **principal components transform**. Instead of deriving the coefficients of the transformation statistically typical values which correspond to attributes of land cover ('brightness', 'greenness; and 'wetness') are used

TBC
To Be Confirmed

temporal compositing
the averaging of values in adjacent members of a multitemporal image set (typically AVHRR) in order to reduce the effects of cloud cover and atmospheric variation

TES
Tropospheric Emission Spectrometer (EOS)

texture
generic term applied to a number of statistical measures of the variability between a pixel and its neighbours

thermal infrared (TIR)
the portion of the electromagnetic spectrum between 8.0 and 22.0 μm available for remote sensing

tie-point
a point which can be identified in two different images used for image to image registration. Similar to a **ground control point** except that the true map coordinate need not be known.

TIGER
Terrestrial Initiative in Global Research

TIN
Triangulated Irregular Network

TLA
Three-Letter Acronym

TM
Thematic Mapper (Landsat)

TMR
TOPEX Microwave Radiometer (EOS)

TOMS
Total Ozone Mapping Spectrometer (ADEOS)

topology
the spatial relationship between geographic objects for example 'is contained within', 'is adjacent to'

training sample
representative samples of known feature types to be used as the basis of multispectral **classification**

TREES
TRopical Ecosystem Environment observation by Satellites

TVI
Transformed Vegetation Index

unsupervised classification
multispectral **classification** performed without the use of training samples and based solely on the multivariate statistical distribution of the image pixels (also known as cluster analysis)

unwrapping
in SAR **interferometry** the algorithm used to convert phase differences in the signal into height values

USGS
United States Geological Survey

UTM
Universal Transverse Mercator

UV
UltraViolet

validation
any method used to confirm the accuracy and reliability of a processing chain. Validation may be subdivided into **blunder detection, process validation** and **product validation**

vector format
digital representation of map features as a series of straight-line segments together with associated **attribute data**

vegetation index
any algorithm used to transform image values into a more direct measure of the vegetation content of a pixel. Often based on a comparison of the response in the red portion of the spectrum to that in the near infrared

VSAR
Versatile Synthetic Aperture Radar (ALOS)

VSI
Vegetative Sponge Index

windowing
selection of map **vectors** that lie within an image boundary

world reference system (WRS)
the path/row convention used for the designation of image location in terms of the repetitive orbit of a satellite (the path) and the distance along a particular orbit path (the row)

XS
eXtended Spectrum (SPOT)

zero doppler
the point of no frequency shift in SAR processing

Index

WITHDRAWN
BOURNEMOUTH UNIVERSITY LIBRARY